Design Center Stuttgart

des Landesgewerbeamts
Baden-Württemberg

Design Center
Stuttgart

des Landesgewerbeamts
Baden-Württemberg

**Internationaler Designpreis
Baden-Württemberg 2002
und Mia Seeger Preis 2002**

**Baden-Württemberg
International Design Award 2002
and Mia Seeger Prize 2002**

Focus Lebensart
Focus way of life

avedition

Inhalt *Contents*

PANAMA
TEST IT.

Dr. Walter Döring – Wirtschaftsminister und Stellvertretender Ministerpräsident Baden-Württembergs
Dr. Walter Döring *– Minister of Economic Affairs and Technology, and Deputy Minister President of Baden-Württemberg*

Lebensart ... Wirtschaft *Way of life ... the economy*

Thema *Topic*

Lebensart – das geht jeden an. Das Leben wird durch den Umgang mit Dingen bestimmt, die wir benötigen. Diese Produkte müssen neben der reinen Funktionalität Faktoren wie Qualität, visuelle Anmutung, Haptik und vieles mehr aufweisen. Schwerpunkt dieses Jahr sind eben jene „weichen" Faktoren. Dadurch erhalten diejenigen Branchen, die bei den Themen der letzten Jahre kaum teilnehmen konnten, eine neue Chance. 30 Prozent mehr Einreichungen verdeutlichen das Interesse und belegen die hervorragende Positionierung des Internationalen Designpreises Baden-Württemberg.

Way of life is something that concerns all of us. Our lives are determined by our dealings with the things we need. Apart from being purely functional, these products must have properties such as quality, visual attractiveness, pleasantness to the touch, and many more. This year's focus is on just such "soft" factors. Because of this, industries which have barely had any possibility to participate given the topics chosen in recent years have been given a fresh chance. A 30 per cent rise in the number of entries is evidence of the interest in the competition, and documents the excellent status enjoyed by the Baden-Württemberg International Design Award.

Innovation und Interaktion *Innovation and Interaction*

Lebensart – die zahlreichen Facetten sind nicht auf einen Nenner zu bringen. Der individuellen Art zu leben wird vielfach Ausdruck mittels individueller Produkte verliehen. Die Wirtschaft ist gefordert, dafür neuartige Lösungen zu suchen und Produkte zu entwickeln, die dazu beitragen, den bevorzugten Lebensstil zu kultivieren und die Qualität zu erhöhen. Funktionales, innovatives und emotionales Design ist gefragt. Design ist wichtig und nicht nur schmückendes Beiwerk. Gutes Design kommt an, macht Spaß und stärkt die wirtschaftliche Schlagkraft der Unternehmen.

Way of life – the many facets of this topic cannot be reduced to one common denominator. In many cases, the way one individual chooses to live is given expression with the help of individual products. The challenge for the world of business is to find innovative solutions for this, and to develop products that help the preferred lifestyle to be cultivated as well as improving quality. Functional, innovative und emotional design is what is needed. Design is important, not just decorative ornamentation. Good design is appealing, is fun, and strengthens companies' economic leverage.

Dr. Friedrich Bullinger – Präsident des Landesgewerbeamts Baden-Württemberg
***Dr. Friedrich Bullinger** – President of the Baden-Württemberg Office for the Promotion of Trade and Industry*

Wirtschaft...Wettbewerb *The economy...competition*

Wirtschaftsförderung qua Wettbewerb *Promotion of trade and industry qua competition*

Focus Lebensart ist ein ausgezeichnetes Instrument der Wirtschaftsförderung, insbesondere für den Mittelstand – schließlich ermöglichen gerade die nicht-kommerzielle Ausrichtung und die günstigen Anmeldungsgebühren auch kleinen und mittleren Unternehmen die Teilnahme. Als Landesgewerbeamt können wir so größenbedingte Nachteile ausgleichen: Die prämierten Unternehmen gewinnen an Publizität und Profil, die Markteinführung der Produkte fällt leichter.

Focus way of life is an excellent tool for the promotion of trade and industry, especially for small and medium-sized enterprises – in the end, its non-commercial character and reasonable registration fee are precisely what allow small and medium-sized enterprises to participate. As the State Office for the Promotion of Trade and Industry we are thus in a position to compensate for the disadvantages of small size: the companies that win awards enhance their reputation and image, and it is easier for them to launch their products.

Wettbewerbsprofil *Competition image*

Die Seriosität des Wettbewerbs steht an erster Stelle und ist für uns absolut verbindlich. So dürfen zum Beispiel Produkte, an deren Gestaltung die JurorInnen mitgewirkt haben, nicht am Wettbewerb teilnehmen. Die Bewertung der Produkte erfolgt ausschließlich im Original oder am Prototyp, dessen Serienreife gewährleistet ist. Eine selektive Foto-Jury kommt vorab nicht in Frage, denn sie könnte zu leicht zu voreiligen und damit falschen Ergebnissen führen.

The respectability of the competition is our prime concern, and is absolutely imperative for us. For example, if any of the judges have been involved in the design of a product, it cannot be entered for the competition. Products are assessed either exclusively in their original form or as prototypes whose readiness for series production is guaranteed. There is no question of any pre-selection on the basis of photographs, as this could easily lead to premature, and thus false, results.

Henning Horn – Leiter des Design Center Stuttgart des Landesgewerbeamts Baden-Württemberg
Henning Horn *– Head of the Design Center Stuttgart at the Baden-Württemberg Office for the Promotion of Trade and Industry*

Wettbewerb...Design *Competition...design*

Designinterpretation *Design interpretation*

Lebensart ist der Motor, und Design ist sein Treibstoff – oder umgekehrt? Beide sind untrennbar verbunden, wenn es darum geht, unseren „Way of Life" zu gestalten. Damit meine ich gewiß nicht nur die Spaßseite des Lebens, die Welt von Genuß, Schönsein und augenfälligem Luxus. Der Alltag ist es, mit Arbeit, Familie, Freizeit und Erholung, der nach Produkten verlangt, die praktisch und leicht zu handhaben sind, dabei vielleicht sogar Spaß bereiten, und die damit unsere Lebensqualität verbessern. Ich zweifle daran, daß die meisten Designer „Lebensart" bewußt als Thema aufgreifen. Sie entwerfen zielgerichtet für ausgewählte Marktsegmente und Käufergruppen. Nur aktuelle und attraktive Entwürfe bringen auf gesättigten Märkten Erfolg. Lebensart entsteht zunächst fast nebenbei, fixiert wird sie erst durch die individuell erworbene Warenmixtur.

Way of life is the motor, and design is its fuel – or is it the other way around? The two are inextricably bound up with each other if the issue is one of giving form to our "way of life". By that, I am by no means thinking of the fun aspects of life, the world of enjoyment, of being beautiful, of ostentatious luxury. It is everyday life, with its work, family, leisure and relaxation, that calls out for products that are practical and easy to use, and that may even be fun in the process, and that improve the quality of our lives in this way. I doubt that most designers deliberately choose "way of life" as a topic. They design specifically for selected market segments and consumer groups. Only up to date, attractive designs can be successful on saturated markets. Initially, way of life arises almost as a sideline, only becoming defined by the mixture of goods that the individual chooses to buy.

Lifestyle?! *Lifestyle?!*

Bei „Lebensart" denkt man zunächst an Lifestyle und der Begriff ist ja bei vielen professionellen Gestaltern in Deutschland nicht gerade positiv besetzt, ganz im Gegensatz zu den englischsprachigen. Genau aus diesem Grund lautet unser englischer Wettbewerbstitel „Focus Way of Life". Darum geht es nämlich: Produkte, die uns helfen, das tägliche Leben zu meistern, und dabei gleichzeitig dem Bedürfnis nachkommen, uns selbst darzustellen. Denn „Lebensart" zeigen, heißt Profil zeigen, und das ist individuell ...

The first thing we associate with "way of life" is lifestyle: a term that does not exactly have positive connotations for many professional designers in Germany, quite unlike their English-speaking counterparts. It is precisely for this reason that "Focus way of life" was chosen as the English title for our exhibition. For this is what we are after: products that help us to master our everyday lives, while at the same time taking account of our need to tell our own story. Presenting one's "way of life" means standing out from the rest, and that is what individuality is about ...

Peter Wippermann – Trendbüro Hamburg

Keine Lebensart ohne Lifestyle

Der Antrieb zum Design ist die Steigerung der Lebensqualität, das Ergebnis ein vielfältiges und inzwischen hochspezialisiertes Puzzle, das unsere Lebenskultur widerspiegelt. Somit ist Design ein wesentliches Ausdrucksmerkmal unserer Lebensart.

Wohlüberlegt hat sich das Design Center Stuttgart für das diesjährige Jahrbuch zum Internationalen Designpreis Baden-Württemberg für den Begriff Lebensart entschieden und selbst in der englischen Übersetzung den Begriff vermieden, der allzuschnell modisch und despektierlich wirkt: Lifestyle. Lebensart distanziert sich von Lifestyle. Gratuliert man einem Designer zu seiner neuen Produktentwicklung, daß er damit genau im Lifestyle-Trend liege, hat man selbst direkt ins Fettnäpfchen getroffen. Mit fünfzigprozentiger Sicherheit wendet sich der Designer ab, mit siebzigprozentiger Sicherheit meidet er jeden folgenden Wortwechsel, mit hundertprozentiger Sicherheit fühlt sich der Designer falsch verstanden. Ging es ihm doch gerade nicht darum, noch ein weiteres Produkt auf den Markt zu werfen, das mit zig anderen auf den gerade aktuellen Lifestyle-Trend aufsetzt. Sind doch vielmehr die Verbesserung der Lebensqualität und damit auch Langlebigkeit sein Ziel, um ein höheres Niveau der Lebensart zu erreichen.

Doch wenn es im Design immer noch um die Verbesserung der Lebenskultur geht, dann ist Lifestyle heute einer der wichtigsten Faktoren, die unsere Lebensart bestimmen. Das folgende Essay beschreibt, warum der Lebensart ohne Lifestyle die Bilder fehlen und warum Design, so es denn noch als Dienstleistung verstanden wird, im Lifestyle mehr Potenzial hat, als es bisher wahrgenommen wurde.

> „Gratuliert man einem Designer zu seiner neuen Produktentwicklung, daß er damit genau im Lifestyle-Trend liege, hat man selbst direkt ins Fettnäpfchen getroffen."

Lange galten Langlebigkeit, Selbsterklärung von Funktion, Produktehrlichkeit und Nutzerorientierung als die Gestaltungsziele der Designer – frei nach der Sullivanschen Formel „form follows function". Dann rief jemand tief aus dem Schwäbischen „design follows emotion" herauf und riß die Designgilde in die Sinnkrise. Der Urfrog Hartmut Esslinger löste die Form von ihrem Inhalt, um sie als kommunikatives Produktinterface dem Konsumenten zuzuwenden. Daß hier Teilbereiche zum hauptsächlichen Inhalt erhoben wurden, widerstrebt bis heute vielen Designern, die einen ideologisch geprägten ganzheitlichen Ansatz verfolgen.

Doch es kam schlimmer: die Ganzheitlichkeit verbuchte ein aufstrebender Bonvivant namens Tyler Brulé für sich, indem er das Designmagazin „Wallpaper" auf den Markt brachte, das Design nicht in seinen Disziplinen, sondern inszeniert in Lebenssituationen zeigte. Seine Ganzheitlichkeit lag in der Verknüpfung von Mode-, Möbel-, food-, packaging-, transportation-, industrial-, consumer-electronic- und interiordesign samt Architektur. Das einzelne Objekt ist nichts, die Inszenierung alles. Er reduzierte Design auf die Oberfläche, degradierte Design zur Mode, inszenierte es in überzogenen Stilisierungen, denen er durch ironisch-versnobte Kommentare den letzten dünnen Boden entzog – und war damit ein besserer Beobachter der allgemeinen Befindlichkeit und der beste Dienstleister in der Branche der Sehnsuchtsstiller. Für die globale Stilgruppe der jungen Jet-Sets mit Hunger nach Extravaganzen, Eigeninszenierung und Stilbildung kreierte Brulé den stylistischsten Lifestyle den es je gab. Sein Erfolg macht klar, daß Lifestyle eben kein beliebiger „stuff that surrounds us" ist, sondern eine wohlgestaltete zusammenhängende Tapete, ein Stilkondensat, das dem Bedürfnis nach Selbstdarstellung, Halt und Orientierung nachkommt. Lifestyle ist damit der Zeitgeist-Hintergund, vor dem die jeweilige Stilgruppe ihre Lebensart erst ausleben kann.

Peter Wippermann – Trendbüro Hamburg

No way of life without lifestyle

Design is motivated by the desire to improve the quality of life, and its result is a multi-facetted (and meanwhile highly specialized) jigsaw puzzle that reflects our living culture. This makes design a significant expression of the way we live.

It was not without good cause that the Design Center Stuttgart chose "way of life" as its motif for this year's Baden-Württemberg International Design Award Yearbook, deliberately avoiding the term "lifestyle", which in German can all too quickly seem voguish and pejorative. Way of life has set itself apart from lifestyle. In Germany, it is a faux pas to congratulate a designer on his new product development by saying that he has had an eye for precisely the right lifestyle trend. There is a fifty-fifty chance the designer will turn away, seventy percent certainty that he will avoid any further conversation, and absolute certainty that the designer will feel misunderstood. After all, the whole point of his endeavour was not to launch yet another product on the market that tried to joint the umpteen others on the latest lifestyle/trend bandwagon. Rather, his aim is an improvement in the quality of life and thus also longevity, in order to achieve a way of life on a higher level.

And yet, if design is still concerned with improving living culture, then lifestyle is currently one of the most important factors shaping our way of life. The following essay describes why way of life (as an attitude) without lifestyle (as representation) lacks visualization, and why design, if it is still understood as a service, has more potential in lifestyle than German designers have so far realized.

"In Germany, it is a faux pas to congratulate a designer on his new product development by saying that he has had an eye for precisely the right lifestyle trend."

For many years, longevity, self-explanatory function, product honesty and user orientation were seen as the goals of any designer's creativity – loosely following Sullivan's dictum "form follows function". Then someone from the depths of southern Germany invoked the formula "design follows emotion", and plunged design and designers into a crisis of identity. Hartmut Esslinger, the "granddaddy frog", separated form from its content in order to present form to the consumer as a communicative product interface. Even today, this elevation of parts' status to become the main content is a bone of contention for many designers who pursue an ideologically conditioned, holistic approach.

Yet there was worse to come: an up-and-coming bon vivant called Tyler Brulé credited himself with a holistic approach by launching the design magazine "Wallpaper". This did not show design in its disciplines but presented it in the stage setting of life situations. Its holistic approach consisted in combining fashion, furniture, food, packing, transportation, industrial, consumer electronic and interior design, including architecture. The individual object is nothing, its setting everything. He reduced design to surface, demoted design to fashion, placed it in hyperbolic settings, which he pushed over the edge into deconstruction with cynical, snobbish commentaries – and in doing so was a better observer of general sensibilities and the best service provider in the desire-satisfying industry. For the global style group of the young jet sets yearning for extravagance, to be noticed, and to have a style of its own, Brulé created the most style-conscious lifestyle there had ever been. His success is clear evidence that lifestyle is not simply the chance "stuff that surrounds us", but well-designed, matching wallpaper, an essence of style that satisfies the need for self-presentation, security and guidance. Lifestyle is thus the zeitgeist, the background against which the various style groups can define their way of life.

Wirkung statt Wirklichkeit

Ein weiterer Aspekt, der Lifestyle als Orientierungsmerkmal geradezu notwendig macht, ist die zunehmende Produktkomplexität. Früher erklärte sich die Funktionsweise von gut gestalteten Produkten an ihren nach außen hin sichtbaren Wirkungszusammenhängen von selbst. Heute kann man nur noch die Wirkung, den Output auf einen bestimmten Input registrieren, aber nicht mehr den Inhalt „begreifen", da er sich entmaterialisiert, digitalisiert, miniaturisiert oder anderweitig aus dem Staub der Wahrnehmbarkeit gemacht hat.

Der eigentliche Mechanismus verbirgt sich im Innern einer Blackbox. Wir haben uns damit abgefunden, die dingliche Umwelt weder verstehen noch begreifen zu können und uns damit den wenigen Spezialisten hoffnungslos ausgeliefert. Wir hangeln uns mit einer „try and error"-Taktik an den Abgründen einer technisierten Welt entlang, die wir nicht einmal mehr partiell verstehen.

„Wir haben uns damit abgefunden, die dingliche Umwelt weder verstehen noch begreifen zu können und uns damit den wenigen Spezialisten hoffnungslos ausgeliefert."

Unser Denken wird dabei in eine lineare Struktur gezwungen, die auf dem ständigen Nachvollziehen der primitiven Logik der Maschinen beruht. Unsere Handlungsweise wird von den Konstrukteuren unserer objektuellen Umwelt vorgedacht, während die weitaus mächtigeren und genialeren Fähigkeiten unseres intuitiven, kreativen Gehirns verkümmern, weil sie weder vom User dieser Technologien benutzt werden können, noch sich vom Designer auf die einfältige aber unglaublich schnelle Maschine übertragen lassen. Dadurch bleibt meist eine tiefe Frustration im Umgang mit diesen Objekten zurück.

Wir übertragen diesen Mechanismus von Input/Output auf unseren Lebensstil. Lebensziel ist es nicht mehr, die Wirklichkeit zu verstehen. Entscheidend wird die Frage, mit welchem minimalen Aufwand wir eine bestimmte Wirkung erzeugen können, um ein gewünschtes Ziel zu erreichen. So werden wir immer mehr zu Menschen, denen es nicht mehr wichtig ist, grundsätzlich zu verstehen, warum wir uns so oder so fühlen, sondern unser Selbstbild davon ableiten, wie wir auf andere wirken und diese auf uns reagieren. Die Chance zur Selbstverwirklichung wird dadurch zwar verwirkt, die Chance auf Selbstdesign dagegen immer wieder gegeben. Unser Leben gleicht immer mehr einer durchsichtigen Skeleton-Swatchuhr, in deren Gehäuse sich zwar sichtbar noch ein paar Zahnräder bewegen, aber niemand mehr zu beurteilen vermag, ob sie überhaupt noch zur Funktion beitragen oder nur der Dekoration dienen, um die Anmutung einer Uhr und von vergehender Zeit zu erwecken.

Nutzerorientiertheit

Wie soll nun aber ein Designer das gestalten, was auch ihm einerseits unverständlich ist und andererseits nicht die kreativen Möglichkeiten eröffnet, die er für eine am Menschen orientierte Gestaltung bräuchte? Die Lösung sind künstliche Ersatzwelten, die es erlauben, seelische und körperliche Bedürfnisse zu befriedigen oder auszuleben, die im Berufs- und Alltagsleben nicht mehr benötigt oder angesprochen werden. Für die ständig Unterforderten wird die Welt überfüllt mit multimedialen Erlebnisinseln, virtuellen Abenteuerwelten und Soziotainmentprodukten, für die permanent Überforderten mit Relaxstationen, Wellnessfarmen und anderen Fluchtressorts, in die bei Bedarf abgetaucht werden kann, um das innere Gleichgewicht wiederzuerlangen.

Effect instead of materiality

One further aspect that makes lifestyle almost necessary as a benchmark is products' increasing complexity. With their externally visible chain of cause and effect, it used to be self-explanatory how well designed products worked. Today, all we can do is register effect, how output follows a certain input, but we can no longer "grasp" content, because it has been dematerialized, digitalized, miniaturized or otherwise been transported from the realms of perceptibility. The actual mechanism is concealed inside a black box. We have resigned ourselves to being unable either to understand or grasp our tangible environment, and have put ourselves completely at the mercy of the few specialists there are. By trial and error, we grope our way along the precipice of a technological world that we have not even begun to understand.

As we do so, our thoughts are forced into a linear structure based on permanent aping of machines' primitive logic. Our actions are predetermined for us by the developers of our objectual environment, while the far more powerful and genial abilities of our intuitive, creative brain shrink, either because they cannot be used by the person using these technologies, or because designers cannot transfer them to the simple, but incredibly fast, machine. The usual result is profound frustration when dealing with these objects.

> *"We have resigned ourselves to being unable either to understand or grasp our tangible environment, and have put ourselves completely at the mercy of the few specialists there are."*

Let us transfer this input/output mechanism to our lifestyle. It is no longer our aim in life to understand reality. The decisive issue now is that of using minimum effort to generate a specific effect, in order to attain a desired goal. We are thus increasingly becoming people who no longer care about understanding why we feel the way we do, but who derive their self-image from the way we act on others and they react to us. While this means that we forfeit any chance of self-realization, it allows us the opportunity to design ourselves again and again. Our lives are increasingly coming to resemble a transparent Skeleton Swatch, where a few cogwheels can be seen moving, but where nobody can tell any more whether they have any real function or serve solely as decoration to suggest the idea of a watch and of passing time.

User orientation

How, then, should a designer go about designing something that, on the one hand, is incomprehensible even for him and, on the other, does not allow him the creative possibilities that he would need for an anthropocentric design? The answer lies in artificial surrogate worlds which allow those intellectual and physical needs to be satisfied or lived out that are no longer needed or addressed at work or in everyday life. For people who are constantly under-challenged, the world is packed full of multimedia islands of experience, virtual adventure worlds and sociotainment products, while for those who are permanently over-challenged there are places to chill out, beauty farms and other places of refuge, where the visitor can get away from it all if he needs to, in order to regain his inner balance.

Ein Beispiel ist der Trend zum virtuellen Freund, automatisierten Haustieren wie AIBO oder Q-Taro, dem neuesten Produkt aus dem Sonystall. Q-Taro ist ein kugelförmiger „Healing"-Roboter der Wärmesensoren besitzt und der mit Licht- und Klangeffekten den psychischen Zustand des Besitzers beeinflussen soll. Bewußt wurde auch hier versucht, die nachgewiesenen heilenden Aspekte von Haustieren auf eine Maschine zu übertragen, ohne die Nachteile eines lebendigen Tieres in Kauf nehmen zu müssen. Dieses Bündel aus Plastik und Schaltkreisen macht weder Schmutz noch verlangt es Nahrung, Auslauf oder Zuneigung, wenn sich Frauchen/Herrchen auf einer längeren Dienstreise befindet. So wird sich die Aufgabe der Designer in Zukunft weniger damit befassen, leblose Produkte zu entwerfen und diesen einen formalen Charakter zu verleihen, sondern lebendig erscheinenden und eigenwillig agierenden Objekten eigenständige Persönlichkeiten zu verleihen, die sich an die Mentalität und temporären Launen des jeweiligen Benutzers anpassen können ohne ihn zu belasten.

Die Objektwelt übernimmt mehr und mehr die Funktion der Lieblingspuppe eines Kindes, dessen Eltern gerade nicht da sind. Der von Einsamkeit und Leere geplagte Single erhält von seinen Produkten das tröstliche Gefühl vermittelt, von vielen treuen Freunden umgeben zu sein. Dies gibt ihm scheinbaren Halt in einer haltlosen Welt. So könnten diese automatisierten Artgenossen bald zu einer neuen Lebensart führen.

Früher genügte es, wenn Produkte nur sich selbst und ihre Funktion möglichst einfach erklärten. Heute müssen sie mit einem Besitzer kommunizieren, der es verlernt hat oder unfähig ist, sich selbst auszudrücken. Die Ursachen dafür sind vielfältig. Neonomadische Lebensführung und damit verbundene Fernanwesenheit, Zeit- und Leistungsdruck und nicht zuletzt fehlendes Selbstbewußtsein in einer medialisierten Welt, deren von professionellen Werbestrategen perfekt konstruierten Idole zu übermächtig und unerreichbar für Normalsterbliche sind. Produkte dienen heute dem Besitzer als Persönlichkeitsviagra, das die Kaste der Designer in Form von immer neuen Lifestyleprodukten, Markenartikeln und anderen Wichtigkeit verleihender Gadgets zu liefern hat.

„Produkte dienen heute dem Besitzer als Persönlichkeitsviagra, das die Kaste der Designer in Form von Wichtigkeit verleihenden Gadgets zu liefern hat."

Vom Lifestyle zum temporären rent-a-style

Längst sprechen wir im Trendbüro nicht mehr von demografischen Zielgruppen, sondern von Stilgruppen, einer Art Community of choice (Wahlgemeinschaft) mit stilistischem Konsens. Nicht die Altersgrenzen clustern mehr die unterschiedlichen Weltbilder. Vielmehr sind es individuelle Idealvorstellungen, die mittels wirtschaftlicher Leistungsfähigkeit als eine temporäre Identität geborgt und eklektizistisch zusammengekauft werden. Menschen – und längst nicht mehr nur Japaner, deren persönliche Individualität kulturell durch extreme Sozialisation weniger ausgeprägt ist – legen sich mit Hilfe eines Lifestyles eine Persönlichkeitsfassade zu, einen „Designcharakter", der aus einem der vielen Trendmagazine herausgesucht wird. Lifestyle wird damit zur konsumierbaren Lebensart. Ein so geborgter Lebensstil bietet den immensen Vorteil, daß er nicht verteidigt werden muß. Sollte es irgendwann zu viel Energie erfordern, ein Image aufrechtzuerhalten, wird das temporäre Selbstbild einfach wie eine Maske gewechselt.

One example is the trend towards the virtual friend, to automatic pets such as AIBO or Q-Taro, the latest product from the Sony stables. Q-Taro is a spherical "healing" robot. Equipped with heat sensors, it is supposed to use light and sound effects to affect its owner's mental state. A conscious attempt has also been made here to transfer the proven healing aspects of pets to a machine, without having to accept the disadvantages of a living animal. This assemblage of plastic and integrated circuits neither makes a mess nor does it demand to be fed, to be taken for a walk or to be given affection while its owner is away on business for a few days. In future, then, the designer's task will be less that of designing inanimate products and giving them a formal character, and more one of giving personalities of their own to objects that appear animate and act with a mind of their own, and that can adapt to the mentality and passing moods of their owners without causing them any trouble.

"Today's products are a kind of Viagra for their owner's personality, which the designer caste has to provide in the form of gadgets bestowing importance."

The world of objects is increasingly assuming the function of the favourite teddy bear of a child whose parents have just gone out for a while. These products will give the singleton stricken by loneliness and emptiness the comforting feeling of being surrounded by many faithful friends. This will give him the semblance of a fixed centre in a world whose centre cannot hold. Automatic creatures such as these could soon result in a new way of life.

It used to be enough if products were simply self-explanatory and their function could be understood as easily as possible. Today, they have to communicate with an owner who has forgotten how to express himself, or is incapable of doing so. There are many reasons why this is so, such as a neo-nomadic lifestyle and the state of being virtually present while absent that goes with it, the pressure of time and work and, not least, the lack of self-confidence in a media-dominated world, whose idols, perfectly constructed by professional advertising strategists, are too overbearing and unattainable for normal mortals. Today's products are a kind of Viagra for their owner's personality, which the designer caste has to provide in the form of ever new lifestyle products, brand articles and other gadgets bestowing importance.

From lifestyle to temporary rent-a-style

At the Trendbüro, we have long since stopped speaking about demographic target groups, preferring instead to think in terms of style groups, of a kind of community of choice that adheres to a consensus in matters of style. The various ways of seeing the world no longer form clusters along age barriers, but rather, a temporary identity is borrowed and purchased eclectically, according to individual ideals and purchasing power. People – and this by no means applies any longer solely to the Japanese, whose personal individuality is culturally less pronounced due to their extreme socialization – take on a personality façade with the help of a lifestyle, a "designer character" selected from one of the many trend magazines. Lifestyle is thus becoming a way of life that can be consumed. A lifestyle that has been borrowed in this way has the considerable advantage that it does not have to be defended. If it should at some time become too strenuous to keep up an image, the temporary image the person has chosen for himself will simply be exchanged like a mask.

Peter Wippermann – Trendbüro Hamburg

> „Die Mediengesellschaft im Optimierungsdruck überprüft sich ständig neu und adaptiert neue Stil- und Wertemuster wie ein Chamäleon."

War Langlebigkeit immer ein Hauptanliegen von Design, ist sie in der heutigen Wirtschaft eher ein unerwünschter Nebeneffekt – wenngleich kein offen eingestandener. Verkürzte Produktzyklen und steigender Preisverfall unterstützen den Wandel, daß Konsumgüter, die bisher durch ihren materiellen Wert und ihre Funktionalität zeitbeständig waren, zu Mode- oder gar Wegwerfprodukten werden. Die Mediengesellschaft im Optimierungsdruck überprüft sich ständig neu und adaptiert neue Stil- und Wertemuster wie ein Chamäleon. War früher Lebensstil etwas, das sich in irgendeiner Form durch das ganze Leben gezogen hat, ist es heute ein temporärer Durchgangszustand. Dies hat Auswirkungen auf unsere Lebensart, denn dadurch wandelt sich auch unsere Lebenskultur. Nicht mit jeder Mode und auch nicht mit gleicher Geschwindigkeit, doch wenn Werte sich wandeln, ändern sich damit auch unsere Definitionen von Lebensqualität. Unsere Art zu leben ist keineswegs trendimmun, sondern reagiert zwangsläufig auf die neuen Anforderungen von Mobilität, Schnelligkeit und Flexibilisierung. Der Lifestyle selbst wird zum Greifstyle, den man zwar nicht begreift aber sich einfach greift – zur Be- oder Entschleunigung –, wie und wenn er gerade benötigt wird. Unser Leben entwickelt sich zur ästhetischen Entscheidung.

Design ist Service im Markt der Gefühle

Lifestyle als Greifstyle bedeutet, daß die nötigen Codes, Regeln und Kommunikationsformen innerhalb der gesellschaftlichen Gruppen leicht erlernt und adaptiert werden können, um Reibung bei einem Wechsel zu vermeiden. Dies läßt sich einfacher über Lifestyleprodukte vermitteln, die eine bestimmte Zugehörigkeit wortlos kommunizieren können, als über geistige Haltungen. Ich habe das Handy X, also gehöre ich zur Gruppe X. Daß es mittlerweile nicht nur in Japan üblich ist, mehrere Mobiltelefone gleichzeitig zu besitzen, nicht nur um die verschiedenen Lebensbereiche wie Arbeit, Familie und Lover säuberlich getrennt halten zu können, sollte in diesem Zusammenhang nicht unerwähnt bleiben.
Wie Swatch einst die passende Uhr zum Outfit erfunden hat, so liefern die Designer in den Elektronikkonzernen in immer kürzeren Abständen die neuen paßgenauen Bausteine für die Szene im persönlichen Lebensfilm, der gerade inszeniert werden soll. Die Objekte werden zur Dekoration unseres Seelenzimmers, zum Tapetenmuster unseres sozialen Aufenthaltsraums. Produkte werden deshalb auch immer weniger mit ihren eigenen, kaum mehr unterscheidbaren Qualitäten beworben, sondern über ein konstruiertes holistisches Lebensgefühl verkauft. Wie der Marlboromann die Freiheit und nicht das Rauchen verkauft, so erwirbt man heute mit Kauf eines Produktes eine ganze Lebensphilosophie, ein persönliches Corporate Identity.

Das Produkt selbst wird unwichtig und austauschbar und ist nur noch durch sein Äußeres differenzierbar, welches perfekt mit dem zu transportierenden Image korrelieren soll. Beim Sony Ericsson Handy erleichtern auswechselbare Gehäuseabdeckungen das temporäre Switchen von einem Lebensstil in einen anderen.

Letztendlich ist es bei der zunehmenden Komplexität in Funktionalität und der Bedienung von Produkten auch kaum mehr sinnvoll, die Inhalte auszutauschen. Die User wollen sich nicht jedes Mal mit einer neuen Bedienphilosophie auseinandersetzen, nur weil sie den Anbieter oder das Modell wechseln. Software und Interfacestrategien werden firmenübergreifend und weltweit standardisiert, um schnelles Erlernen zu ermöglichen oder ganz überflüssig zu machen. Die Hardware selbst wird von nur wenigen Herstellern für fast alle Anbieter hergestellt. Die Bindung an ein Produkt kann deshalb nur noch über emotionale Ebenen hergestellt werden, da es nur noch wenige technische und funktionelle Unterscheidungsmerkmale gibt.

Peter Wippermann – Trendbüro Hamburg

While longevity was always one of design's main concerns, it tends to be an undesired side-effect in the modern-day economy – even though this may not be admitted openly. Shortened product cycles and the accelerating decline in prices are bringing about a situation where consumer goods, whose material value and functionality have so far made them durable, are becoming fashion items, or indeed disposable products. In the race to be better and better, the media society is constantly reappraising itself, adapting new patterns of style and value like a chameleon. Lifestyle used to be something that was present in one form or another over an entire lifespan, but today it has become a temporary, transient condition.

"In the race to be better and better, the media society is constantly reappraising itself, adapting new patterns of style and value like a chameleon."

This has repercussions for our way of life, because our living culture is changing as a result. Not with every fashion, nor at the same speed, but if values change then this also involves a change in our definitions of quality of life. Our way of living is by no means immune to trends, but is bound to react to the new demands of mobility, speed and flexibility. Lifestyle itself is becoming a pick 'n' mix commodity, which is grabbed without being grasped – to speed up or slow down – as and when it is needed. Our life is becoming an aesthetic decision.

Design is service in the emotion market

Pick 'n' mix lifestyle means that the codes, rules and communication forms of social groups can easily be learnt and adapted in order to avoid friction when changing to a new one. It is easier to communicate this via lifestyle products, which can convey affiliation to a specific group non-verbally, than by intellectual attitude. I own mobile X, therefore I belong to group X. In this context, it should not be forgotten that it has since become common, and not only in Japan, to own several mobile telephones at once, and not only in order to be able to keep the various areas of life such as work, family and lover neatly separated.

Just as Swatch once invented watches to match outfits, designers in electronics companies are ever more frequently providing the new perfectly matching props for those scenes in personal biographies that are about to be shot. Objects are becoming the ornaments of our "soul rooms", the wallpaper pattern of our social waiting room. This is why products are being promoted less and less for their inherent qualities, which scarcely distinguish them from others, and instead being sold via a construed, holistic feeling of life. Just as the Marlboro man sells liberty rather than smoking, the person buying a product today acquires a whole philosophy of life, a personal corporate identity.

The product itself is becoming insignificant and interchangeable, only distinguishable from others by virtue of its exterior, which should correlate perfectly with the image to be conveyed. In the case of the Sony Ericsson mobile, exchangeable housing covers make it easier to switch temporarily from one lifestyle to another.

In the end, as the function and operation of products becomes ever more complex, it scarcely makes any more sense to exchange contents. Users do not want to have to come to terms time and again with a new mode of operation simply because they have changed their provider or model. Software and interface strategies are being standardized across companies and across the world in order to allow rapid learning, or indeed to make new learning superfluous. Hardware itself is produced by only a few manufacturers for almost all providers. The only way loyalty to a product can therefore be created is via the emotional level, given that there are only few technical and functional distinguishing features left.

Dadurch aber wird die Zunft der Designer gerade zu dem degradiert, was sie nicht sein will: Verpackungsgestalter und Stylisten, die emotional aufgeladene Hüllen entwerfen und sich schlimmstenfalls ewig in der Retro-Schleife drehen. Dennoch muß der Lifestyletrend nicht gleich zum Verfall der Berufsethik führen. Unternehmer und Marketingstrategen wissen heute, daß die Kommunikation schon beim Produktdesign beginnen muß, und räumen der Designzunft daher zunehmend mehr Einfluß ein. Um den Mangel an Innovationen zu kompensieren, erfinden Designer mit den Marketingstrategen Geschichten, schlagen Brücken zu Sehnsüchten, kreieren Mythen – kurz: montieren einen Lifestyle, der mehr erzählt, als es der Produktnutzen und das Produkt allein je vermocht hätten. Diese Überhöhung ist trotz ihres naturgemäßen Erscheinens an der Oberfläche keine beschränkte Oberflächlichkeit, sondern eher eine neu geknüpfte Beziehung zwischen dem Konsumenten, seinen Sehnsüchten und Wunschwelten.

Lifestyle ist nicht zuletzt die clevere Methode, durch stilisierte Bilder beim Konsumenten immer neue Wunschwelten zu wecken und die Bausteine für seinen Lebensstil gleich mit zu liefern. Lifestyle ist damit auch der Baustein zur Lebensart. Die Steigerung der Lebensqualität bedeutet heute also nicht mehr primär Langlebigkeit und Funktionalität. Im Zeitalter der Medialisierung und Individualisierung bedeutet Lebensqualität Orientierung, Gefühl und Vertrauen. Es ist ein Trugschluß, zu glauben, man könne Lebensart losgelöst von Lifestyle und Zeitgeist praktizieren. Selbst die Abkehr vom Trend ist eine Reaktion auf den Trend.

Peter Wippermann
Trendbüro Hamburg

For those in the design profession, this means they are being degraded to something they do not want to be: packaging designers and stylists designing emotionally laden shells, at worst perpetually caught up in the retro loop. Nevertheless, the lifestyle trend must not automatically mean the demise of our professional ethos. Today's businessmen and marketing strategists know that communication has to start right from the design of a product, and they are therefore allowing the design profession ever greater influence. To compensate for the lack of innovation, designers are working together with marketing strategists to invent stories, to forge links with desires, to create myths – in short, to put together a lifestyle, that tells more than product benefit and the product alone could ever have done. Although this "added value" must by its very nature be a surface phenomenon, it is no trifling bagatelle, but rather a newly forged relationship between the consumer, his desires and his dream worlds.

Lifestyle is not least a clever method of using stylized images to arouse ever new dream worlds in the consumer, while at the same time furnishing the building blocks for his lifestyle. Lifestyle is thus also a building block towards a way of life. Today, therefore, improving the quality of life is no longer primarily a matter of longevity and functionality. In the age of medialization and individualization, quality of life is a matter of orientation, emotion and trust. We are deceiving ourselves if we believe we can practice our way of life divorced from lifestyle and zeitgeist. After all, to turn one's back on a trend is to react to that trend.

Peter Wippermann
Trendbüro Hamburg

Focus Lebensart – Der Wettbewerb

Vielfach drücken sich individuelle Lebensweisen, -stile oder gesellschaftliche Anschauungen in bestimmten Gebrauchsgegenständen aus. Produkte können nicht nur zur Imagebildung ihres Nutzers beitragen, sondern auch zu neuen Verhaltensweisen führen. So hat beispielsweise das Kickboard mit großem Erfolg eine andere Spielart der Fortbewegung hervorgerufen: die Mikromobilität. Umgekehrt werden neue Lebensstile von Herstellern aufgegriffen und fließen in die Gestaltung von Produkten ein. Beispielsweise wirken sich veränderte Eßgewohnheiten, wie etwa die Vorliebe für die mediterrane Küche oder auch für asiatische Gerichte, wie Sushi, auf Entwürfe von Geschirr und Besteck aus. So beeinflußt unsere Art zu leben die Produktgestaltung und umgekehrt haben Produkte wiederum Einfluß auf unsere Lebensart.

So wurde dieses Jahr der Wettbewerb unter das Thema Focus Lebensart gestellt. Der Internationale Designpreis Baden-Württemberg wird jährlich als Staatspreis für zukunftsweisende und herausragende Gestaltungsleistungen an Hersteller und Designer verliehen. Seit geraumer Zeit wird er unter einem Thema ausgeschrieben, das jährlich wechselt – eine besondere Herausforderung für die Einreicher. Gebrauchsgegenstände sollen zur Verbesserung des Lebens im Alltag, bei der Arbeit, in der Freizeit, im Urlaub beitragen. Produkte unter dem Focus Lebensart gehen über die reine Funktionalität noch hinaus. Visuelle Anmutung, Haptik und andere „weiche", emotionale Faktoren bestimmen die Freude und den Spaß an einem Produkt, was bei der Bewertung der Einreichungen in dem Wettbewerb eine große Rolle spielte.

Focus in Gold – Focus in Silber

In den vergangenen Jahren erhielten die Einreicher für höchste Designleistungen den „Internationalen Designpreis Baden-Württemberg" und für gute Designleistungen das Label „Ausgezeichnet! – Design Center Stuttgart". Dies führte immer wieder zu Unklarheiten, die man mit neuen Auszeichnungen beseitigen möchte. So erhalten gut gestaltete Produkte seit diesem Jahr den „Focus in Silber", die besten davon – pro Kategorie jedoch nur ein Produkt – den „Focus in Gold". Neu ist auch, daß der Mia Seeger Preis 2002 unter dem gleichen Thema „Focus Lebensart" ausgeschrieben wurde.

Firmen und Designer konnten unter folgenden Bedingungen Produkte für den Wettbewerb „Focus Lebensart" einreichen:

- Der Hersteller oder der Designer von industriellen Serienprodukten mußte Sitz, Niederlassung oder Vertretung in Deutschland haben.
- Das Produkt durfte nicht länger als zwei Jahre auf dem Markt sein.
- Zugelassen waren auch Prototypen, deren Serienreife gewährleistet war, jedoch keine Unikate.
- Produkte der Juroren konnten nicht am Wettbewerb teilnehmen.

Außerdem sollten die eingereichten Produkte einer der folgenden 12 Kategorien entsprechen:

- Wohnen (Möbel, Inneneinrichtungen...)
- Textilien (Teppiche, Gardinen, Stoffe...)
- Küche, Sanitär (Küchen- und Badezimmereinrichtungen, Sanitärprodukte...)
- Haushalt (Haushaltsutensilien, Töpfe, Geschirr, Besteck, elektrische Haushaltsartikel...)
- Outdoor (Gartenartikel, Reisebedarf, Camping...)
- Sport, Spiel, Wellness, Freizeit (Sportgeräte und Zubehör, Spielzeug, Saunazubehör...)
- Büroausstattung, Accessoires (Schreibgeräte und Zubehör...)
- Objektmöblierungen (Büro-, Lounge-, Countermöbel...)
- Beleuchtung (Lampen, Leuchten...)
- Architektur (Architekturelemente, Gebäudekonzeptionen...)
- Transport, Verkehr (Autos, Motorräder, sonstige Verkehrsmittel...)
- Kommunikation, Information, Foto, Optik (elektronische Produkte, Media-Produkte, Kameras, Brillen...)

Focus way of life – The competition

In many instances, individual ways of life, lifestyles or social viewpoints are expressed in certain objects of everyday use. Not only can products contribute towards the image of their user, they can also lead to new modes of behaviour. The kickboard, for example, has been extremely successful in bringing forward a new form of locomotion: micromobility. Conversely, new lifestyles are picked up by manufacturers and find their way into the design of products. Changed eating habits, for example, such as the craze for Mediterranean cuisine or for Asian dishes such as sushi, affect the way crockery and cutlery are designed. In this way, our way of life influences product design, and products in their turn influence our way of life.

The topic chosen for this year's competition, then, was focus way of life. The Baden-Württemberg International Design Award is given annually to manufacturers and designers as a government prize donated by the state of Baden-Württemberg for forward-looking and outstanding design work. For some time now, the competition has been centred around a theme, which changed from year to year – quite a challenge for entrants. Objects of everyday use should contribute to improving everyday life, working life, leisure time or holidays. To qualify for focus way of life, products had to go beyond mere functionality. Visual attractiveness, pleasantness to the touch and other "soft", emotional factors determine whether a product will be a pleasure or fun, and this played a great role in the assessment of the entries for the competition.

Focus in Gold – Focus in Silver

In past years, the competitors presenting the best design work were given the "Baden-Württemberg International Design Award", while good designs were given the label "Outstanding! – Design Center Stuttgart". Again and again, this resulted in confusion, and the new awards are intended to remedy this. This year, well designed products receive for the first time the "Focus in Silver", while the best products – but only one product per category – receive the "Focus in Gold". A further new feature is that entries were also invited for the Mia Seeger Prize 2002 under the same "Focus way of life" theme.

Companies and designers were allowed to enter products for the competition on the following conditions:

- *The competition was open to any manufacturer or designer of industrial series-manufactured products domiciled in Germany, or with a subsidiary or branch office there.*
- *Any product could be entered which had not been on the German market for longer than two years.*
- *Prototypes whose readiness for series production was guaranteed could also be entered; one-off products could not compete, however.*
- *The judges' own products were barred from the competition.*

The products submitted had to fit into one of the following 12 categories:

- *Home (furniture, interiors…)*
- *Textiles (carpets, curtains, fabrics…)*
- *Kitchen, sanitary installations (kitchen and bathroom fittings, sanitary products…)*
- *Household (household appliances, pots and pans, crockery, cutlery, electrical household items…)*
- *Outdoor (products for the garden, travel requisites, camping…)*
- *Sport, recreation, wellness, leisure (sports equipment and accessories, toys, sauna accessories…)*
- *Office equipment, accessories (writing implements and accessories…)*
- *Contract furniture (office, lobby, counter furniture…)*
- *Lighting (lamps, luminaires…)*
- *Architecture (architectural elements, building design…)*
- *Transport, traffic (cars, motorcycles, other means of transport…)*
- *Communication, information, photography, optical equipment (electronic products, media products, cameras, spectacles…)*

226 Enreicher aus aller Welt meldeten insgesamt 430 Produkte an. Das sind 30 Prozent mehr Einreichungen. Damit stieß der Wettbewerb unter dem Thema Focus Lebensart auf ein noch größeres Interesse als im Vorjahr, was die rege internationale Beteiligung noch unterstrich. So kamen ausländische Designer aus Italien, England, Australien, Vietnam, den USA, Japan, Österreich, Norwegen, Finnland, Frankreich, Belgien, Dänemark, den Niederlanden und der Schweiz. Bei den Herstellern waren Länder vertreten wie Italien, Spanien, England, Dänemark, Schweden, Norwegen, Finnland, Belgien, Australien, Japan, Taiwan, Nepal, den USA, den Niederlanden, Österreich und der Schweiz.

Die Auswahl

Eine international erfahrene und anerkannte Expertenjury kam für zwei Tage in Stuttgart zusammen, sichtete, befühlte, prüfte, diskutierte und bewertete die Originalprodukte, bis sie zur endgültigen Auswahl kam. Dabei legte sie folgende Bewertungskriterien zugrunde:

- Gestaltungsqualität
- Innovationsgehalt
- Ergonomie
- Verwendung neuer Werkstoffe
- ökologischer Aspekt
- Gebrauchsqualität
- Benutzerführung, Produktgrafik beziehungsweise -typographie
- gestalterische Qualität der Produktausstattung: Bedienungs-, Montageanleitung, Verkaufsverpackung
- Wertigkeit, Anmutung und emotionaler Gehalt

Aus den 430 eingereichten Produkten wählten die Juroren 83 aus, die sie für auszeichnungswürdig hielten. 75 mal vergaben sie dabei den „Focus in Silber" und acht mal den „Focus in Gold". Alle 83 Produkte sowie die Wettbewerbsergebnisse des Mia Seeger Preises sind vom 10. Oktober bis zum 8. Dezember 2002 im Haus der Wirtschaft zu sehen. Die acht Produkte, die den „Focus in Gold" gewonnen haben, werden zusätzlich auf den Internetseiten des Design Center Stuttgart präsentiert und auf Pickup-Karten veröffentlicht. Besucher der Ausstellung wählen ihren persönlichen Produktfavoriten und ermitteln damit den Publikumspreis, der am Ende der Ausstellung verliehen wird.

Orelisierte Preisträger

Die mit dem „Focus in Gold" ausgezeichneten Produkte wurden von dem Kreativstudio Frank M. Orel für Werbe- und Modefotografie interpretiert und fotografisch umgesetzt. Der Stuttgarter und sein achtköpfiges Team haben ihre Linse statt auf das Produkt auf das Thema „Leben" gerichtet. Sie wollten keine Werbe- und Profimodefotografie. So entstanden zumeist aus dem Leben gegriffene Bilder, die nicht nur Wärme und Leben(digkeit) ausstrahlen sondern auch Geschichten erzählen und den Zeitgeist einfangen. „Es sollte so aussehen, wie wenn alles im Vorübergehen geknipst worden wäre. Dem ist aber nicht so, es sind überall Gedanken dahinter. Es wurde ja inszeniert, was bei manchen Produkten nicht ganz einfach war", erklärt Frank M. Orel das fotografische Konzept. „Für das normale Leben sind die Produkte nur Attribute. Sie sollen ja dem Menschen dienen und nicht umgekehrt." So stellte er das Produkt häufig in unerwartete Zusammenhänge, in denen es eher nebensächlich wirkt. Einige Fotos wurden geradezu zu Suchbildern, bei denen der Betrachter Spaß daran hat, das Objekt erst auf den zweiten Blick zu entdecken. Außerdem wurde „orelisiert", das heißt mit den verschiedensten Methoden verfremdet, einmontiert, gespiegelt, ungewöhnliche Perspektiven auf die Produkte eröffnet. Frank M. Orel und sein Team setzten die Fotos für „Focus Lebensart" als Reportage um, jedoch mit einigen gewollten Sprüngen. „Denn das Leben hat ja auch Sprünge", so Frank M. Orel. Etwas verbindet dennoch alle Bilder, etwas Emotionales.

226 entrants from all over the world submitted a total of 430 products: an increase of 30 percent. The "Focus way of life" topic thus met with an even greater response than last year's topic, and this was further highlighted by the great number of international entries. There were foreign designers from Italy, the UK, Australia, Vietnam, the USA, Japan, Austria, Norway, Finland, France, Belgium, Denmark, the Netherlands and Switzerland. Among the manufacturers, the countries represented included Italy, Spain, the UK, Denmark, Sweden, Norway, Finland, Belgium, Australia, Japan, Taiwan, Nepal, the USA, the Netherlands, Austria and Switzerland.

The selection process

For two days, a panel of internationally experienced and recognized experts met in Stuttgart, viewed, felt, examined, discussed and assessed the original products, until they reached their final selection. Their assessment was based on the following criteria:

- design quality
- innovative content
- ergonomics
- use of new materials
- ecological aspect
- quality of use
- user controls, product graphics and typography
- design quality of product peripherals: instructions for use and assembly, sales packaging
- quality, grace and emotional content

From the 430 products submitted, the judges chose 83 that they felt were worthy of commendation. Of these, 75 received the "Focus in Silver", and eight were given the "Focus in Gold". All 83 products selected and the results of the competition for the Mia Seeger Prize will be exhibited in the Haus der Wirtschaft from 10 October to 8 December 2002. In addition, the eight products awarded the "Focus in Gold" will be presented on the Design Center Stuttgart website and published on pick-up cards. Visitors to the exhibition will choose their personal favourite from the products on show and in this way vote for the winner of the "Popular Award", which is awarded at the end of the exhibition.

Orel-ized award-winners

The products awarded the "Focus in Gold" have been interpreted and given photographic expression by the Frank M. Orel advertising and fashion photography studio. Stuttgart-based Frank M. Orel and his eight-man team directed their lenses at the topic of "life" rather than at the product. They did not want to produce advertising or professional fashion photography. In most cases, the result was pictures from real life that not only exude a sense of warmth and vitality, but also tell stories and capture the zeitgeist. "The idea was to make it look as though everything had been photographed in passing. This was not the case, however, all of them are based on an idea. They have been stage-managed, even though this was not so easy for some of the products", says Frank M. Orel, explaining his photographic concept. "In normal life, the products are only attributes. After all, they should serve man, not vice versa". Frequently, therefore, he placed the product in unexpected contexts, in which its effect is more one of triviality. Some of the photographs have become veritable puzzles, where the beholder can enjoy only discovering the object at second glance. In addition, the photographs have been "Orel-ized", with a wide variety of methods being used to alienate, superimpose, reflect and open up unusual perspectives on the products. Frank M. Orel and his team gave their photographs for "Focus way of life" the form of a report – a report that suddenly and deliberately changes the subject now and then. "Life itself is full of leaps and bounds", is Frank M. Orel's comment. Nevertheless, there is an emotional quality linking all the pictures.

„Ein Regal muß nicht immer nur im Wohnzimmer oder Büro stehen." Harley-Davidson-Werkstatt Stuttgart; Metallregal „Steel 12"

"A shelf does not always have to stand in a living room or office." Harley-Davidson garage, Stuttgart; "Steel 12" metal shelving

„Die Backofen-Kombination ist heiß, der Mann, der ihn bedient, ist noch heißer." Backofen-Kombination „BDGL 8664-2 UT"

"The oven combination is hot, the man operating it hotter still." "BDGL 8664-2 UT" combination oven

„Ist es nicht eine Freude, mit diesem Löffel zu essen?" Fütterlöffel „Babybjörn® Feeding spoon"

"Isn't it a joy to eat with this spoon?" "Babybjörn®" feeding spoon

„Hier wäre jeder gerne Pferd!" Pferdestall auf dem toskanischen Landgut der Contessa Conti; Reithelm „Duell"

„Who wouldn't love to be a horse here!" Stables on the Tuscan estate of Contessa Conti; „Duell" riding cap

„Im ersten Moment muß man sich kurz Gedanken machen…, was ist das? Ah ein Stuhl. Sehr dekorativ." Armlehnstuhl/Loungechair „DE 773"

"At first you have to ask yourself… what's that? Oh yes, a chair. Very decorative." "DE 773" armchair/lounge chair

„Er hat eine für seine Zwecke gut eingerichtete Leseecke." Antiquitätenladen in Stuttgart; Tischleuchte „Soon"

"He has a reading corner that is well-stocked for his purposes." Antique shop in Stuttgart; "Soon" table lamp

„Die Raststätte ist jenseits von gut und böse, spiegelt aber Zeitgeschichte und Kultur wider" Raststätte am Fernpass; Zeitschriftenhalter „Redondo"

"This service station is beyond good and evil, but it also reflects contemporary history and culture." Service station on the Fern Pass; "Redondo" magazine rack

„Es ist jedes Mal ein Erlebnis, wenn man zu Rosella und Maurizio geht." Ferrari-Stammlokal „Montana" in Maranello; Heizkreisregelung „Vitotronic"

"Every time you visit Rosella and Maurizio, it's a real experience." Ferrari regulars' spot "Montana" in Maranello; "Vitotronic" heating circuit control

LED ZEPPELIN

Focus in Gold

Hersteller ***Manufacturer***	**Vertrieb** ***Distributor***	**Design** ***Designer***
Behindertenwerkstatt Reinsdorf gemeinnützige GmbH Gabelsberger Straße 8 D-08141 Reinsdorf Tel. +49 3 75 2 77 59 22	Behindertenwerkstatt Reinsdorf gemeinnützige GmbH Gabelsberger Straße 8 D-08141 Reinsdorf Tel. +49 3 75 2 77 59 22	Rüdiger Schaack Teichstraße 30 D-08289 Schneeberg

Metallregal „Steel 12" *metal shelving*

Dank seiner schlichten und zeitlosen Gestaltung läßt sich „Steel 12" vielseitig und flexibel einsetzen und fügt sich in jede Umgebung harmonisch ein: als Raumteiler, zur Präsentation von Waren, zur Aufbewahrung von Büchern, Ordnern oder Schallplatten, Badutensilien, Geschirr und vieles mehr. Außerdem lassen sich mittels Magneten Zettel und Fotos anbringen. Das hochwertige Regal aus pulverbeschichtetem Stahlblech in Steckbauweise ist das Ergebnis einer Zusammenarbeit der Fachschule für Angewandte Kunst Schneeberg und der Behindertenwerkstatt Reinsdorf auf der Basis der in der Werkstatt vorhandenen Technik. Es besteht aus vier Stangen, zwei bis fünf Einlegeböden und doppelten Seitenteilen, die um 90 Grad versetzt auf die Stangen geschoben werden und für die Standfestigkeit sorgen. Die Einlegeböden lassen sich einfach auf die Seitenteile auflegen. Das ganze Regal wird werkzeuglos montiert. Es ist in den Farben tiefschwarz, verkehrsrot, graualuminium und perlweiß sowie auf Wunsch in weiteren Farben erhältlich, wobei auch innerhalb eines Regales verschiedene Farben eingesetzt werden können.

The plain, timeless design of "Steel 12" means that it can be used flexibly and in a variety of ways, and that it will harmonize with any type of surrounding: as a partitioning element, to display goods, to store books, folders or records, bathroom articles, china, and much more. In addition, magnets can be used to attach notes and photographs. This high-quality shelving is made of powder-coated sheet metal and features plug-in assembly. It is the result of collaboration between the Schneeberg College of Applied Art and the Reinsdorf Disabled Persons' Workshop, and uses the technology already on hand in the workshop. It comprises four columns, between two and five shelves and double-walled side units, which are slid onto the columns at ninety degrees to each other and provide a solid support. The shelves can easily be rested on the side units. The entire shelf unit is assembled without any tools. It is available in the colours deep black, traffic red, grey-aluminium and pearl, as well as in other colours on request, and different coloured elements can be used in one and the same shelf unit.

Jury ***Judges panel***

Im Kontext von Hightech-Produkten eine verblüffend einfache Lösung. Arretiert einfach durch die Materialstärke. Schöne Farbwahl, schöne einfache Materialien. Hervorragendes Beispiel für Manufakturgedanken, das heißt mit Einfachstmaschinen ein komplexes System zu machen. Besticht durch enorme Bescheidenheit und hohe Multifunktionalität.

A surprisingly simple solution in the world of high-tech products. It is held in place simply by the thickness of the material. Good colour scheme and good simple materials. Excellent example of the "workshop" idea, i.e. the idea of making a complex system using the most basic machines. Stands out by virtue of its extremely unassuming nature and highly multifunctional design.

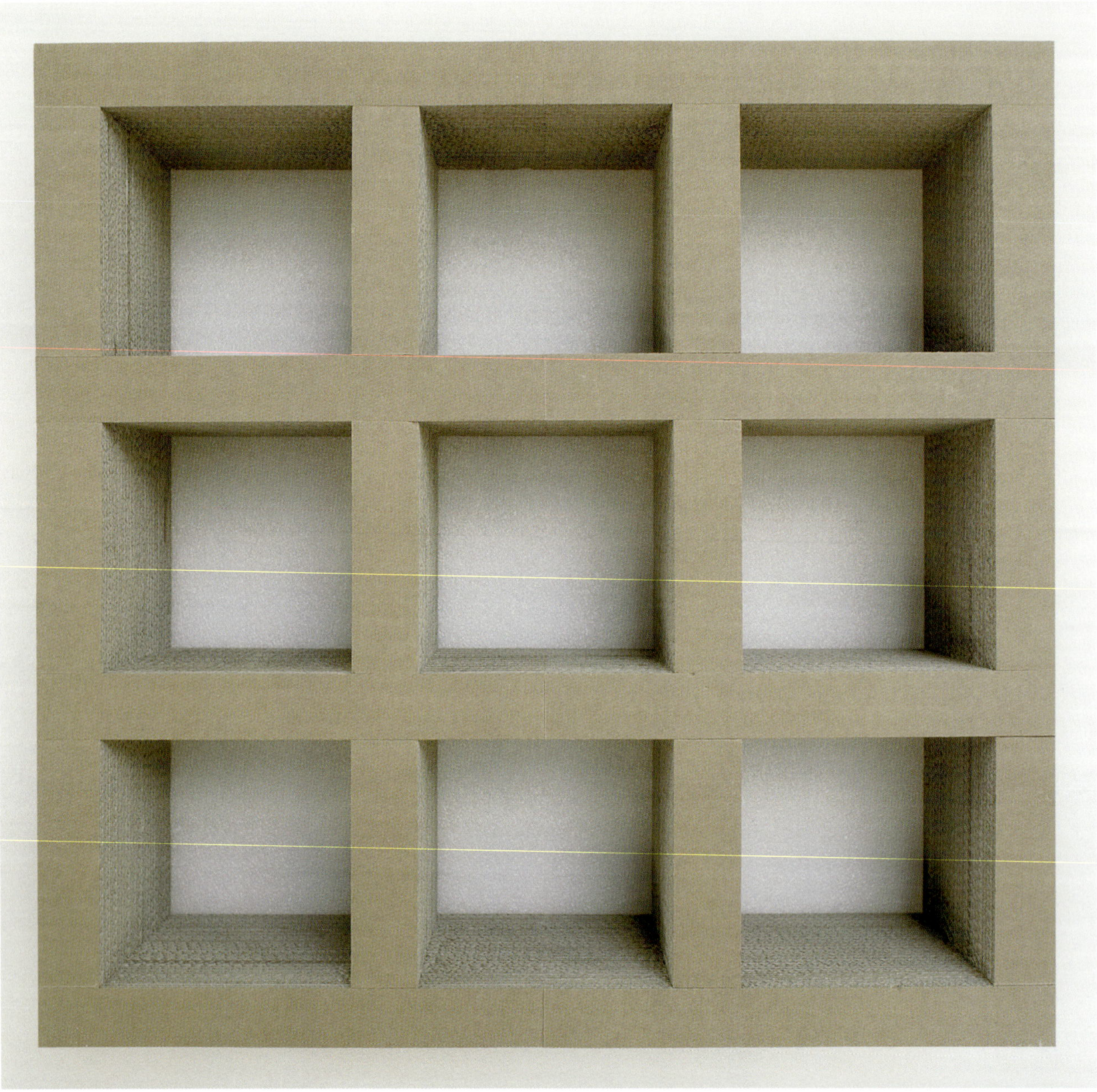

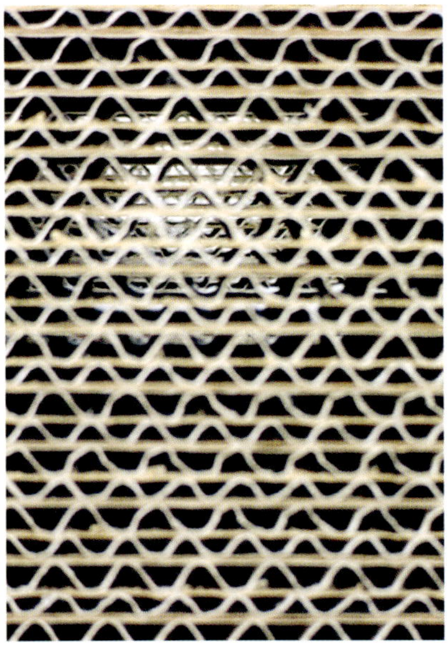

Vertrieb ***Distributor***

Metris Planungsgesellschaft GbR
Rhönring 121
D-64289 Darmstadt
Tel. +49 61 51 99 77 17
www.metris-office.de

Design ***Designer***

Dr. Silke Nuß, Nuno Rodrigues
Mittnachtstraße 1
D-70191 Stuttgart
Tel. +49 1 60 92 51 55 22

CD-Regal *CD shelf*

Außergewöhnliche optische und haptische Eigenschaften zeichnen dieses Regal aus. Dank der offenen Wellenstruktur, durch die man durchschauen kann, wird es zu einem transparenten, licht- und luftdurchlässigen Möbel. Außerdem wirkt die Wellenstruktur wärmedämmend, so daß sich die Flächen nicht nur weich sondern auch warm anfühlen. Neun würfelförmige Fächer bieten Platz für 126 CDs. Um es an der Wand aufzuhängen, wird es lediglich auf zwei Wandschrauben in den oberen Ecken aufgelegt. Das Möbel besticht durch Leichtigkeit, unerwartete Langlebigkeit und hohe Stabilität sowie unglaubliche Tragfähigkeit. Die offenporige Oberfläche verhindert eine statische Aufladung der Flächen, das bedeutet Staub wird nicht angezogen. Das CD-Regal wird folgendermaßen gefertigt: Aus der Wellpappe werden Streifen in Laufrichtung der Wellen geschnitten, so daß die Wellen sichtbar sind. Anschließend werden die Streifen geschichtet und miteinander verklebt. Um Ecken oder Stöße auszubilden, versetzt man die Streifen nach jeder Schicht gegeneinander. Das Regal ist kompostierbar, da es nur aus Wellpappe und Stärkeleim besteht.

This shelf stands out by virtue of its extraordinary visual appearance and feel. Because its open, undulating structure can be seen through, it is a transparent piece of furniture through which light and air can pass. The undulatory structure also has an insulating effect, so that its surfaces are both soft and warm to the touch. Nine cube-like compartments provide room for 126 CDs. To mount it on a wall, it is simply hung from two wall screws in the top corners. The fascinating thing about this item of furniture is that it is light, unexpectedly long-lasting, very robust and able to take an incredibly heavy load. The porous surface prevents the shelves from becoming statically charged, and so they do not attract dust. The CD shelf is manufactured like this: strips of corrugated cardboard are cut in the direction of the corrugations, with the result that the corrugations are visible. The strips are then layered and bonded to each other. To form corners or joints, the strips are overlapped after each layer. The shelf is compostable, as it is only made of corrugated cardboard and starch glue.

Jury ***Judges panel***

Schöne Materialwahl, Fetisch Pappe, gute Funktion. Einfache technische Lösung durch simple Stanzungen auch im Sinne des Schönwerdens und des Alterns, relativ unproblematisch also in Abnutzung und Nutzung dieses Regals. Erstaunliche Architekturwirkung.

Nice choice of material, cardboard fetish, works well. Simple technical solution with simple punched perforations, also an allusion to becoming beautiful and growing old, and so wearing out and using this piece of furniture is relatively unproblematic. Astonishingly architectural impression.

Hersteller ***Manufacturer***	**Vertrieb** ***Distributor***	**Design** ***Designer***
Jahn Design Fact Schlesische Straße 27 D-10997 Berlin Tel. +49 30 5 30 00 60 www.jahn-design.com	Jahn Design Fact Schlesische Straße 27 D-10997 Berlin Tel. +49 30 5 30 00 60 www.jahn-design.com	Jahn Design Fact Schlesische Straße 27 D-10997 Berlin Tel. +49 30 5 30 00 60 www.jahn-design.com

Steckregalsystem „Cubico 2" *plug-in shelving system*

„Cubico 2" ist ein einfaches Steckregalsystem, mit dem man nach individuellen Wünschen jedes Regal bauen kann. Es läßt in horizontaler wie in vertikaler Richtung beliebig viele quadratische Fächer zu, die auf dem Quadratraster von 35 Zentimeter basieren. In der Senkrechten wie in der Waagrechten haben sämtliche Regalelemente die gleiche Bauart. Sie sind in regelmäßigen Abständen bis zur Hälfte eingeschlitzt und werden ineinander gesteckt. Es gibt sie in Längen von 70 Zentimeter mit zwei Schlitzen bis 245 Zentimeter mit 7 Schlitzen. Durch zusätzliche V-förmige Nuten erhält das Regalsystem eine sehr hohe Stabilität. Die lineare Verbindung der Elemente ermöglicht eine unbegrenzte Ausdehnung des Regals. Die Überstände der Elemente, die einem halben Raster entsprechen, dienen der Längsverbindung, sie werden über Verbindungsstifte fixiert. Außerdem kann es durch aufgelegte Glasplatten zum Sideboard werden. Die Regalelemente sind aus zwölf Millimeter dicken Birke-Multiplexplatten mit CNC-Technik gefräst, deren charakteristische Schichtstruktur an den Kanten sichtbar ist, während die Flächen braun oder weiß beschichtet sind.

"Cubico 2" is a simple plug-in shelving system allowing any type of shelf to be put together according to individual wishes. Any number of square compartments based on a 35 cm x 35 cm pattern can be added horizontally or vertically. Vertically and horizontally, all the shelf elements are constructed in the same way. Slits are cut half way into them at regular intervals, and they are slotted into each other. They are available in lengths from 70 centimetres, with two slits, to 245 centimetres, with seven slits. Additional V-shaped grooves give the shelving system a very high degree of stability. The linear connection of the elements means that the shelf can be extended almost endlessly. The projecting ends of the elements form half a basic grid and are used to connect further elements horizontally, which are attached using connecting pins. If a glass panel is placed on top, it can also be converted into a sideboard. Using CNC technology, the shelf elements are milled from 12-millimetre thick birch multiplex panels, whose characteristic layered structure is visible on the edges, while the surfaces are laminated in brown or white.

Jury ***Judges panel***

Flexibles Regalsystem, sinnvoll und funktionierend stabilisiert. Formal strenge und gleichzeitig unglaublich leichte Lösung, die Raum läßt für das Wesentliche, nämlich die Bücher. Einfaches Stecksystem, bei dem keine Gebrauchsanweisung notwendig ist.

Flexible shelving system which has been stabilized in a way that makes sense and that works. Formally strict yet incredibly light solution which leaves room for the most important thing, i.e. the books that go on the shelf. Simple plug-in system, for which user's instructions are unnecessary.

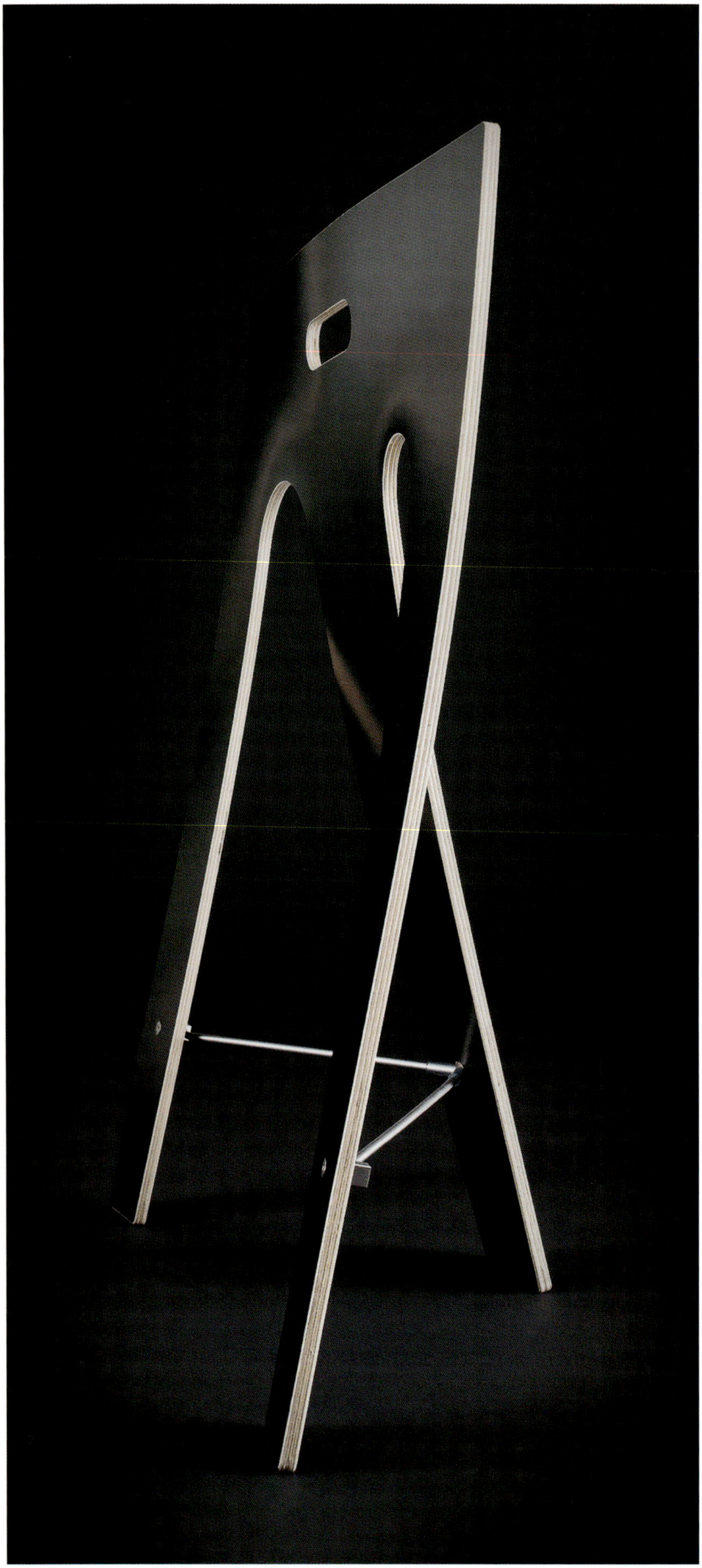

Hersteller ***Manufacturer***	**Vertrieb** ***Distributor***	**Design** ***Designer***
Möbelbau Kaether & Weise GmbH Dammstraße 43 D-31195 Lamspringe Tel. +49 51 83 6 77 www.kaetherundweise.de	Möbelbau Kaether & Weise GmbH Dammstraße 43 D-31195 Lamspringe Tel. +49 51 83 6 77 www.kaetherundweise.de	Thorsten Franck Lautensackstraße 21 D-80687 München Tel. +49 89 57 86 91 94 Fax +49 89 57 86 91 95

Tischbock „Sidestep" *trestle*

Mit zwei Tischböcken und einer Tischplatte läßt sich spielend leicht ein Tisch auf- und abbauen. Benötigt man ihn nicht, kann er platzsparend verstaut und aufbewahrt werden. Seine Stabilität erhält er durch das mittlere Bein, das durch einen eingespannten Edelstahlbügel wie ein Tanzschritt zur Seite – „Sidestep" – abgespreizt wird.
Der Tischbock besteht aus Esche massiv und neun Millimeter filmbeschichtetem Sperrholz. Er ist in den Farben gelb, rot, grün, schwarz, buche, ahorn oder nußbaum zu haben. Neben einer Glasplatte gibt es wahlweise eine Tischplatte in den gleichen Farben wie der Tischbock aus 18 Millimeter filmbeschichtetem Sperrholz.

It is child's play to assemble and dismantle a table that comprises two trestles and a table top. If the table is not needed, it can be stowed away without taking up a lot of space. It gets its stability from its middle leg, which is splayed to the side in a kind of dance step ("sidestep") by a fixed stainless steel brace.
The trestle is made of solid ash and nine-millimetre film-laminated plywood sheet. It is available in yellow, red, green, black, beech, maple or walnut. Apart from a glass table top, there is a choice of table tops made of 18-millimetre film-laminated plywood in the same colours as the trestle.

Jury ***Judges panel***

Zweidimensional wird dreidimensional. Und ein leichter anmutiger Tanzschritt zur Seite. Schöne Laubsägearbeit. Leicht zu verpacken, leicht zu verschicken.

The two-dimensional has become three-dimensional. And a light, graceful dance step to the side. Nice jig-saw work. Easy to pack, easy to dispatch.

Hersteller ***Manufacturer***

Tojo Möbel GmbH
Am Ziegelgraben 26
D-73614 Schorndorf
Tel. +49 71 81 40 14 00
www.tojo.de

Vertrieb ***Distributor***

Tojo Möbel GmbH
Am Ziegelgraben 26
D-73614 Schorndorf
Tel. +49 71 81 40 14 00
www.tojo.de

Design ***Designer***

Roy Schäfer
Weinbergstraße 21
D-75181 Pforzheim
Tel. +49 72 43 33 67 33

Bett „Tojo™ v" *bed*

Ein Bett, das sich den unterschiedlichsten Bedürfnissen und Gegebenheiten anpaßt. Das Jugend-, Azubi-, Studenten- und Singlebett wächst über die Jahre einfach mit und ist dabei äußerst einfach zu handhaben. Wie eine Ziehharmonika läßt sich das Bett stufenlos in der Breite von 90 bis 180 Zentimeter variieren. So kann es auch als Gästebett für eine oder zwei Personen genutzt werden; denn es läßt sich in Windeseile werkzeuglos auf- und abbauen. Benötigt man es nicht mehr, ist es leicht verstaubar. Das Bio-Bett aus unbehandelter Buche Multiplex besitzt weder Schrauben noch Beschläge, weder Lack noch Lasur.

A bed that adjusts to the widest variety of needs and circumstances. This bed for adolescents, apprentices, students and singletons simply grows over the years, and is at the same time extremely simple to use. Like an accordion, the bed can be extended to any width between 90 and 180 centimetres. This means that it can also be used as a spare bed for one or two persons; as it can be assembled and dismantled again in a trice, without any tools. If it is no longer needed, it is easy to stow away. This "bio bed", made of untreated multiplex beech, does not contain any screws or hardware, and is neither painted nor varnished.

Jury ***Judges panel***

Genial für junge Erstausstattung. Bestechendes, ganz einfaches Material. Grafisch interessant. Ganz wichtig die Tradition von Steckprinzipien, wieder Zapfen auszufräsen. Keine Metallverbindungen. Industriell sehr einfache Fertigungsmöglichkeiten. Faszinierend durch die absolute Schlichtheit, ganz knapp am Bastelmoment. Clever.

A brilliant idea for a young person setting up home for the first time. Captivating, extremely simple material. Graphically interesting. One very important feature is the plug-in tradition, cutting out dovetail joints again. No metal connections. Industrially very easy manufacturing possibilities. Its absolute plainness is fascinating, there is almost something do-it-yourself about it. Clever.

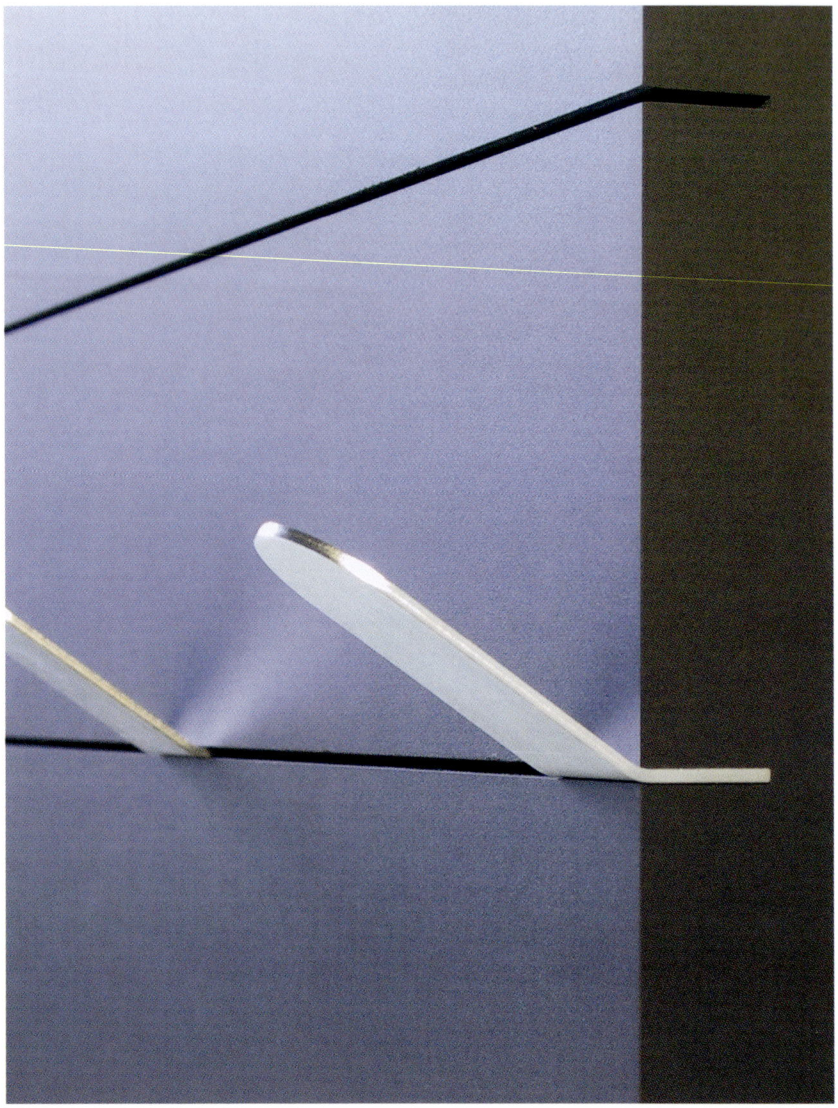

Hersteller ***Manufacturer***	**Vertrieb** ***Distributor***	**Design** ***Designer***
Performa Möbel und Design GmbH Marbacher Straße 54 D-74385 Pleidelsheim Tel. +49 71 44 89 88 69 www.performa.de	Performa Möbel und Design GmbH Marbacher Straße 54 D-74385 Pleidelsheim Tel. +49 71 44 89 88 69 www.performa.de	Martin Kleinhans Werksdesign

Garderobe *coat rack*

Die Garderobe ist das neueste Element einer monolithischen Möbellinie. Wie die besondere Möbellinie selbst ist auch die Wandgarderobe ein extrem grafisches Möbel, das wie eine Farbfläche, ein Displayelement oder ein abstraktes Wandbild wirkt. Die innovative Gestaltung mit Linoleum läßt sie wie einen Monolith erscheinen. Gerade diese Einfachheit verleiht dem auf das Wesentliche reduzierten Möbel seine Wirkungskraft – subtil oder auffällig, je nach Auswahl von Zubehör und Farbe. Dabei kann aus 16 verschiedenen Tönen gewählt werden, die die Wandgarderobe sowohl als auffallende Farbfläche als auch zurückhaltend-still in ein zeitgemäßes Wohn- und Arbeitsambiente integrieren. Das Paneel wird sehr einfach mit Abstand zur Wand montiert. Dadurch entsteht eine Fuge, die den skulpturalen Charakter dieser einfachen und klaren Form unterstreicht. Die Garderobe läßt sich individuell bestücken. Dabei sind die einzelnen Funktionsteile aus Edelstahl, wie Haken und Stange, auf die wesentliche Haltefunktion beschränkt und werden einfach in den scheinbar massiven Linoleumblock eingesteckt.

The hall stand is the latest element in a monolithic line of furniture. Like the special furniture line itself, the wall-mounted coat rack is an extremely graphic piece of furniture, which has the appearance of a coloured surface, of a display element or of an abstract picture. The innovative design using linoleum gives it a monolithic appearance. It is precisely this simplicity that makes this piece of furniture, which has been reduced to the bare essentials, so effective – it can be subtle or striking, depending on the accessories and colour chosen. The choice of 16 different colours means that the coat rack can be either an eye-catching area of colour or, more passively, an integral element of a contemporary living and working environment. The panel is very simply hung, with distance between it and the wall. The resulting gap emphasizes the sculptural character of this simple and clear form. Accessories can be added as desired. The individual functional parts made of stainless steel, such as the hooks and rail, are reduced to their main function as holders, and are simply inserted into the ostensibly solid block of linoleum.

Jury ***Judges panel***

Gegenüberstellung von zwei Einfachstmaterialien, simples technisches Prinzip, formal minimal und puristisch gelöst. Sehr flexibel in der Anwendung. Der Schlitz ist ein Designelement, aber mit Funktion. Und die Garderobe sieht gut aus, wenn sie leer ist. Bei welcher Garderobe ist das sonst der Fall? Ein Beitrag zum mobilen Wohnen, einfach demontabel, leicht und beweglich, veränderbar.

Juxtaposition of two extremely simple materials, simple technical principle, and a formal solution that is minimalist and purist. Very flexible in use. The slit is a design element, but fulfils a function at the same time. And the coat rack looks good when it is empty. What other coat rack can that be said of? A contribution to mobile living, easily taken down, light and movable, changeable.

Hersteller ***Manufacturer***	**Vertrieb** ***Distributor***	**Design** ***Designer***
Walter Knoll AG & Co. KG Bahnhofstraße 25 D-71032 Herrenberg Tel. +49 70 32 20 80 www.walterknoll.de	Walter Knoll AG & Co. KG Bahnhofstraße 25 D-71032 Herrenberg Tel. +49 70 32 20 80 www.walterknoll.de	Pearsonlloyd 42-46 New Road London E1 2AX Tel. +44 20 73 77 05 60 Fax +44 20 73 77 05 50

Sofaprogramm „Easy 480" *sofa series*

„Easy 480" fällt gleich durch seine enorme Größe auf. Es ist eine neuartige, moderne Interpretation eines klassischen Themas – voluminös und filigran zugleich. Das minimalistisch gestaltete Sofa wird charakterisiert durch seinen umlaufenden Rücken und den eingeschobenen Sitz: zwei große Volumen, die sich zu einer Skulptur verbinden. Trotz den auffallend überdimensionierten Maßen, die als bewußtes Stilmittel eingesetzt werden, scheint „Easy" zu schweben dank der filigranen Kufen, die in einem schönen Kontrast zu dem opulenten Sofa stehen. Bei „Easy 480" verbinden sich klare Linien mit rationalen Radien zu einer eigenständigen, geometrischen Sprache. Mit einem einfachen Handgriff läßt sich die großzügige Sitzfläche um 17 Zentimeter nach vorne ziehen, und eine komfortable Liegefläche lädt zum Mittagsschlaf, Hinlümmeln oder Relaxen ein. Überraschend ist dabei auch die veränderte Optik des Sofas. „Easy 480" ist in drei Größen erhältlich: 180, 205 und 230 Zentimeter lang. Dazu gibt es Ecklösungen, Polstersitze und Glastische. Die Sitzecken können mit Anreihsofas, deren Sitz ausziehbar ist, individuell kombiniert werden.

What is immediately striking about "Easy 480" is its huge size. It is a new, modern interpretation of a classical theme – voluminous and filigree to an equal extent. The characteristic features of this minimalistically designed sofa are its wrap-around seat back and the push-in seat: two large masses that combine to form a sculpture. Despite the strikingly oversized dimensions, which are used consciously as a stylistic device, "Easy" appears to float. This is because of its filigree runners, which contrast nicely with the lavish sofa. In "Easy 480", clear lines and rational radii combine to form a unique, geometrical language. In no time at all, the generous seating surface can be pulled forward 17 centimetres, and the result is a comfortable surface on which to doze, sprawl or relax. Another surprising feature here is how this changes the sofa's appearance. "Easy 480" is available in three sizes: 180, 205 and 230 centimetres long. There are also corner seats, upholstered seats and glass tables. The corner seats can be combined individually with extension sofas whose seat can be pulled forward.

Jury ***Judges panel***

Faszinierend, wie mit Übergrößen gespielt wird. Es gelingt, auch mit einer schönen Materiallösung, angenehme großzügige Flächen im Sitzen darzustellen. Leichtigkeit durch die Materialwahl, bestechende Einfachheit der formalen Lösung. Man kann sagen „Back to the future". Funktionsanzeige von der Seite her. Man sieht, daß es ausziehbar ist.

It is fascinating to see how this design plays with excessive size. It manages to present pleasant, generous seating areas, and solves the problem of material choice well. The choice of materials makes it light, the simplicity of the formal solution is captivating. You could say "back to the future". You can see how it works from the side. It is clear that it can be pulled out.

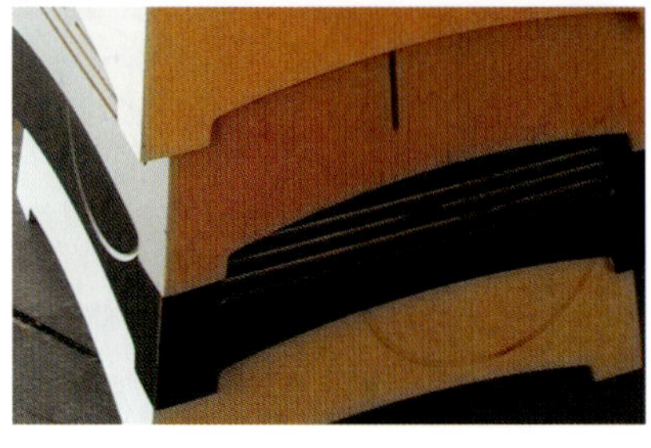

Hersteller ***Manufacturer***	**Vertrieb** ***Distributor***	**Design** ***Designer***
Nils Holger Moormann Möbel-Produktions- und Handels GmbH Festhalle D-83229 Aschau i. Chiemgau Tel. +49 80 52 9 04 50 www.moormann.de	Nils Holger Moormann Möbel-Produktions- und Handels GmbH Festhalle D-83229 Aschau i. Chiemgau Tel. +49 80 52 9 04 50 www.moormann.de	Clemens Stübner/Sabine Mrasek Jakobstraße 90 D-52064 Aachen Tel. +49 2 41 40 02 94 84 Fax +49 2 41 40 02 94 85

Musikhocker „Kleiner Trommler" *percussion stool*

Der „Kleine Trommler" ist Sitzmöbel und Musikinstrument in einem. Seine vier Seiten haben unterschiedlich ausgearbeitete Percussionflächen: Bongo, Congas, Schlitztrommel, Xylo-Trommel und Waschbrett. Bespielt werden die Flächen im Sitzen von Hand oder mit Paukenschlegeln. Das außergewöhnliche Percussionsitzmöbel kann zu Hause, in der Schulpädagogik oder in der Klangtherapie eingesetzt werden. Es ist aus hochwertigem Birkensperrholz und damit wesentlich dimensionsstabiler als Massivholz. Denn hier werden Furniere kreuzweise zusammengeleimt und -gepreßt, was ein Arbeiten des Werkstoffes verringert. Die vier Seitenteile sind zusammen mit der Sitzfläche verleimt und über innenverschraubte Eckleisten zusätzlich verstärkt. Die Percussionflächen werden mit moderner CNC-Technik bis auf eine dünne Schicht eingefräst. Die Größe der einzelnen Flächen bestimmt die jeweilige Tonhöhe. Die eingefrästen Konturen der Percussionflächen verbessern das Schwingungsverhalten des Materials und lassen dank ihres Bildcharakters die jeweilige Funktion besser verstehen. Es gibt ihn lackiert oder schwarzbraun filmbeschichtet. Beide Oberflächen sind sehr langlebig und strapazierfähig.

The "Kleine Trommler" ("little drummer") is a stool and musical instrument in one. Each of its four sides has differently finished percussion surfaces: bongo, congas, slit drum, xylo-drum and washboard. These surfaces are played sitting down, either by hand or with drumsticks. This unusual percussion stool can be used at home, in schoolteaching or in sound therapy. It is made of high quality birch plywood, which makes it far more resistant to warping than solid wood. Here, veneer has been cross-laminated, then glued and pressed together to reduce the likelihood of the material warping. The four side elements have been glued to the seat surface and reinforced with corner battens attached by screws on the inside. CNC technology is used to mill the percussion surfaces down to a thin layer. The size of each surface depends on the pitch that is to be produced. The milled contours of the percussion surfaces help the material to vibrate better, while their pictorial character helps their individual functions to be understood better. It is available either with a varnished finish or with a brown-black film laminate. Both are very long-lasting and hard-wearing.

Jury ***Judges panel***

Das Möbel mit vier Gesichtern. Ganz einfaches Basismöbel, intelligent variiert, das durch Ausfräsungen das Prinzip der Trommel in verschiedenartiger Tonqualität differenziert. Ein vergnügtes Möbel, das zum Spielen einlädt. Es hat eine schöne grafische Qualität.

An item of furniture with four faces. Very simple basic piece of furniture, on which intelligent variations have been made, and which gives different sounds to the drum principle as a result of different depths of milling. A cheerful piece of furniture that calls out to be played with. It has an attractive graphic quality.

12:10
imperial

Hersteller *Manufacturer*	**Design** *Designer*
Imperial-Werke oHG Installstraße 10–18 D-32257 Bünde Tel. +49 52 23 48 12 23 www.imperial.de	T.B.A. Design-Societät Bismarckstraße 50 D-50672 Köln Tel. +49 2 21 5 10 20 03

Backofen-Kombination „BDGL 8664-2 UT" *combination oven*

Das ganz neu gestaltete Gerät ist eine Kombination aus einem Backofen und einem Dampf-Druckgarer. Zehn Backofen-Funktionen verbunden mit der gesündesten Garmethode. Beim Garen mit Dampf, was schon im alten China angewandt wurde, bleiben Geschmack, Aussehen und wertvolle Inhaltsstoffe erhalten. Außerdem können verschiedene Lebensmittel mit unterschiedlichen Garzeiten gleichzeitig gegart werden. Eine Geschmacksübertragung – etwa von Fisch auf Gemüse – ist ausgeschlossen, da immer mit Frischdampf gegart wird. Da der Garraum hermetisch geschlossen ist, bleibt auch die Küche frei von Gerüchen und Schwaden. Bei der Backofen-Kombination verbindet sich hochwertige Technik mit einem außergewöhnlichen Design. Das Gerät zeichnet sich durch seine elegante und edle Erscheinung sowie durch eine klare auf das Wesentliche reduzierte Gestaltung aus: dunkles Anthrazit wird umrahmt von Edelstahl. Das fast schwarze Glas beim Backofen ist undurchsichtig, erst im Betrieb gibt das breite Fenster die Sicht auf den Backraum frei. Die einheitliche Front wird nur durch die zwei langen Türgriffe und das sehr übersichtlich gestaltete Bedienfeld durchbrochen.

This completely re-designed appliance is a combination of an oven and a steam pressure cooker. Ten oven functions combined with the healthiest cooking method. When cooking with steam, a method used in ancient China, taste, appearance and valuable vitamins are preserved. In addition, different foods with different cooking times can be cooked at the same time. Food does not take on the taste of other foods – vegetables do not taste of fish, for example – as fresh steam is always used for cooking. As the cooking vessel is hermetically sealed, the kitchen remains free of odour and steam. In this oven, high-quality technology is combined with an unusual design. The appliance stands out by virtue of its elegant, high-quality appearance, and by virtue of its clear, no-frills design: dark anthracite with a stainless steel surround. The almost black glass of the oven is opaque, and the oven interior is only visible through the wide window when the oven is in operation. The uniform front panel is punctuated only by two long door handles and the extremely clearly designed control panel.

Jury *Judges panel*

Klare und geradlinig gestaltete Front, außerordentlich einfach zu handhabender Dampfkocher. „Two in one". Von außen denkt man, daß es einteilig ist. Vor allem mit nur einer Steuerungsleiste. Viel kleine Kreativität „within the limits". Überzeugend der liegende Dampftopf, da man hier auf verschiedenen Ebenen die Tableaus einbringen kann, also sehr sinnvoll genutzt.

Clear and simply designed front panel, an unusually easy-to-use steam cooker. "Two in one". From the outside, the impression is that of a one-part appliance. Especially as there is only one control bar. A lot of creativity within the limits. The horizontally arranged steam cooker makes a lot of sense, as the trays can be inserted on different levels.

BOSCH
SOLITAIRE

Hersteller *Manufacturer*	**Vertrieb** *Distributor*	**Design** *Designer*
Robert Bosch Hausgeräte GmbH Hochstraße 17 D-81669 München Tel. +49 89 45 90 01 Fax +49 89 45 90 23 47	Robert Bosch Hausgeräte GmbH Hochstraße 17 D-81669 München Tel. +49 89 45 90 01 Fax +49 89 45 90 23 47	Werksdesign

Solitaire-Standherd „Bosch HSW 195 A" *stand-alone cooker*

Mit dem 90 Zentimeter breiten Herd kann man nahezu professionell kochen und backen. Dazu trägt nicht nur der leistungsstarke Wokbrenner in der Mitte bei, sondern vor allem auch das großzügig dimensionierte Kochfeld. Dieser Profi-Rangecooker zeichnet sich durch eine Edelstahlfront, ein Edelstahlgehäuse und hochwertige Gußroste aus. Dank der großen Schalter läßt er sich leicht und gut bedienen. Der Standherd ist funktional und übersichtlich gestaltet. Ins Auge sticht das große Sichtfenster zum Backofen, dessen abgerundete Ecken einen schönen Kontrast zu der rechteckigen Formgebung des Herdes bilden. Auf Wunsch wird der Standherd mit einer Rückwand aus Edelstahl angeboten.

This 90 centimetre-wide cooker allows almost professional standard cooking and baking. This is not only due to the powerful wok burner in the middle, but also to the generously dimensioned hob. The distinguishing features of this professional range cooker are its stainless steel front, stainless steel housing and high-quality cast iron grilles. The large knobs make it easy to use. The stand-alone cooker's design is functional and clear. One striking feature is the large oven window, whose rounded corners form a nice contrast to the rectangular design of the cooker. On request, the stand-alone cooker is available with a stainless steel rear panel.

Jury *Judges panel*

Eine sehr männlich anmutende Koch-Werkstatt. Da hat die Frau in der Küche nichts mehr zu suchen. Da, wo Männer, wenn sie Kochmöbel für Frauen gemacht haben, versagt haben, fängt der Mann plötzlich an, gut zu werden, wenn er für Männer gestaltet. Denn plötzlich ist das ganze Teil auch leicht zu säubern und hat keine Kriechrinnen mehr, in denen sich der Schmutz sammelt.

A cook's workshop with extremely male qualities. There's no place for a woman in a kitchen like this. After being a failure in making cooking appliances for women, men are now suddenly beginning to excel when they design for men. For suddenly, the whole thing is easy to clean, too, and there are no longer any crannies in which dirt can collect.

SIEMENS

Hersteller *Manufacturer*	**Vertrieb** *Distributor*	**Design** *Designer*
Siemens Electrogeräte GmbH Hochstraße 17 D-81669 München Tel. +49 89 45 90 23 87 www.siemens-hausgeraete.de	Siemens Electrogeräte GmbH Hochstraße 17 D-81669 München Tel. +49 89 45 90 01 www.siemens-hausgeraete.de	Frank Rieser, Gerd E. Wilsdorf Werksdesign

Gas-Standherd „Modul-Line Range Cooker HM 19650"
stand-alone gas cooker

Der „Range Cooker" ist ein freistehender großer (90 Zentimeter breiter) Gas-Standherd, mit fünf Kochstellen und einem großzügig dimensionierten Elektro-Backofen ausgestattet. Das sehr große Türfenster gibt einen guten Einblick in den Garraum. Auffallend sind die hochwertigen Gußroste, die nicht über die Herdhöhe hinausragen, sondern auf gleichem Niveau wie die massiven Seitenwangen aus Edelstahl abschließen. Die Edelstahl-Knebel bilden einen schönen ästhetischen Kontrast zur geschliffenen Edelstahloberfläche. Der „Range Cooker" zeichnet sich durch eine klare, sachliche, funktionale Gestaltung aus. Der massive Stangengriff ebenfalls aus Edelstahl paßt zum Siemens Einbauherdprogramm.

The "Range Cooker" is a free-standing, large (90 centimetres wide) gas cooker equipped with five burners and a generously dimensioned electric oven. The very large window in the oven door makes it easy to see what is going on inside. One eye-catching feature is the high-quality cast iron grilles, the top surface of which is flush with the solid stainless steel cheeks. The stainless steel knobs form a nice aesthetic contrast to the ground stainless steel surface. The design of the "Range Cooker" is clear, sober and functional. The solid rail, also made of stainless steel, matches the Siemens built-in oven range.

Jury *Judges panel*

Robustes, klares Design, Bekenntnis zu Edelstahl, konsequent durchgeführt. Schöne Oberfläche, vertiefte Gasdüsen, für kochende Männer. Immer größer, immer mehr Fetisch, deshalb auch mehr Beachtung von Details und von Form, das heißt, das Design kommt durch die Hintertür wieder ganz stark in die Küche. Wir kochen weniger als früher, aber dafür wird es zum Vergnügen.

Robust, clear design, shows a liking for stainless steel, put rigorously into practice. Beautiful surface, sunken gas jets, for men who cook. Ever bigger, ever more fetish, and thus more attention to detail and form. In other words, design is coming back into the kitchen in a big way, but through the back door. We are cooking less than we used to, but when we cook we do it for pleasure.

CERAN
NEFF

Hersteller ***Manufacturer***	**Vertrieb** ***Distributor***	**Design** ***Designer***
Neff GmbH Ruiterstraße 8 D-75015 Bretten Tel. +49 72 52 97 64 04 Fax +49 72 52 97 65 80	B/S/H/ München CNLK – Z Hochstraße 17 D-81669 München	Joachim Grützke Ruiterstraße 8 D-75015 Bretten Tel. +49 72 52 97 63 14 Fax +49 72 52 97 65 30

Glaskeramik-Kochfeld „Twist SE 1389" *ceramic hob*

„Twist SE 1389" macht das Kochfeld zum vollkommen autarken Küchengerät. Auf den vier Kochfeldern der 80 Zentimeter breiten Glaskeramik-Platte mit den verschiedenen Zuschaltzonen gibt es Platz für alle denkbaren Topfkombinationen. Die einzelnen Funktionen werden durch Antippen der integrierten Sensortasten bedient. Die Kochstufe wird über einen Drehwähler auf dem Kochfeld geregelt. Der abnehmbare, nur durch Magnetkraft gehaltene Knebel reguliert alle vier Kochzonen und schaltet sie automatisch aus, sobald er vom Kochfeld entfernt wird – neben der einfachen Reinigung eine ideale Kindersicherung. Das komfortable Kochfeld verfügt zudem über eine elektronische Ankochautomatik, eine automatische Sicherheitsschaltung und einen Timer. Dieser schaltet die Kochfelder nach Ablauf der Zeit automatisch aus oder dient als Kurzzeitwecker. Die Digitalanzeige zeigt neben den Heizstufen der einzelnen Kochzone auch in zwei Stufen die jeweilige Restwärme an, die zum Warmhalten verwendet werden kann. Eine geschwungene Linie trennt das Bedienfeld optisch vom Kochbereich ab.

"Twist SE 1389" makes a completely self-sufficient kitchen appliance out of the hob. On the four cooking zones on this 80 centimetre wide ceramic hob, with its various extension zones that can be additionally switched on, there is room for every conceivable combination of saucepans. The individual functions are controlled by the integrated touch controls. Heat is controlled by a knob on the hob. The removable knob, which is only held in place magnetically, controls all four cooking zones and switches them off automatically as soon as it is removed from the hob – apart from making the hob easy to clean, this is a perfect child locking device. In addition, this easy-to-use hob has an electronic control that automatically reduces power once the desired temperature has been reached, an automatic safety switch and a timer. This either switches off the hob automatically when time is up, or it can be used as a kitchen timer. Apart from the heat selected for the individual cooking zones, the digital display gives a two-stage indication of how much residual heat is left (so that it can be used to keep food warm). A curved line separates the control panel optically from the cooking area.

Jury ***Judges panel***

Bestechend ist dieses Ein- und Ausschalten durch Mitnahme des Switch on and off, was die Gefahr für Kinder reduziert. Die Gefahr ist trotzdem noch da, aber kann von Kindern wenigstens nicht manipuliert werden. Auffallendes, großzügiges Format der Kochplatten. Es gelingt, gleichzeitig einen großen Topf und einen ovalen, etwa einen Fischtopf, aufzusetzen.

One striking feature is how it is switched off simply by taking the off switch away, thus reducing the danger for children. Danger is still there, but it can at least not be manipulated by children. The generous format of the cooker rings catches the eye. It is possible to put a large saucepan and an oval one (like a fish kettle) on the cooker at the same time.

Hersteller *Manufacturer*	**Vertrieb** *Distributor*	**Design** *Designer*
Neff GmbH Ruiterstraße 8 D-75015 Bretten Tel. +49 72 52 97 64 04 Fax +49 72 52 97 65 80	B/S/H/ München CNLK – Z Hochstraße 17 D-81669 München	Ralf Grobleben Ruiterstraße 8 D-75015 Bretten Tel. +49 72 52 97 61 49 Fax +49 72 52 97 65 30

Wandesse „DSL 89" wall mounted extractor hood

Die Wandesse „DSL 89" aus Edelstahl ist die ideale Ergänzung zum Glaskeramik-Kochfeld „Twist SE 1389". Auch sie bietet neben dem geradlinigen, klaren und sachlichen Design Technologie auf höchstem Niveau. Trotz der enormen Absaugleistung ist diese Dunstabzugshaube erstaunlich leise. Die Saugleistung kann in vier Stufen über die Drucktasten auf der Frontseite eingestellt werden. Das große Display und die Leuchtbausteine zeigen die einzelnen Stufen, den Nachlauf und die Filtersättigung an. Die drei leicht herausnehmbaren Metallfettfilter sind so ausgelegt, daß sie hervorragend sowohl den Rand als auch die Fläche absaugen. Somit werden Fettdämpfe, Kochschwaden und -gerüche komplett absorbiert. Drei integrierte Halogenstrahler, die sich separat zuschalten lassen, bieten eine angenehme Arbeitsbeleuchtung. Die schwarze Glasfrontblende nimmt in ihrer geschwungenen Form die Linie, die das Bedienfeld vom Kochfeld „Twist SE 1389" abtrennt, sowie die Optik der Blenden der Neff-Backöfen wieder auf.

The stainless steel "DSL 89" extractor hood is an ideal complement to the "Twist SE 1389" ceramic hob. Like the latter, it combines uncomplicated, clear and sober design with advanced technology. Despite its very powerful extraction fan, it is surprisingly quiet. Buttons on the front allow extraction speed to be adjusted to one of four settings. The large panorama display shows settings, after-running and whether the filter needs changing. The three metal filters, which are easy to remove, have been located in such a way that they excellently extract air from the edge of the cooker, as well as its main surface. Fat, steam and odour are thus completely absorbed. Three integrated and separately switched halogen lights provide a pleasant working light. The concave line on the black glass front panel replicates the line dividing the "Twist SE 1389" control panel from the hob, and conforms with the visual appearance of Neff ovens' front panels.

Jury *Judges panel*

Klare Geometrie, einfaches Auswechseln der Filter, dezente Lichtsteuerung, wo normalerweise Neon-Röhren sind, sind es hier Halogenlampen, also ein für die Wohnküche angenehmes Licht und eine angenehme Atmosphäre. Es scheint so zu sein, daß die kühle Atmosphäre wegbleibt, daß wir hier also ein technologisches Gerät integrieren und uns damit trotzdem wohlfühlen.

Clear geometry, filters are simple to remove, lighting is discreet; where there are normally fluorescent strip lights we have halogen lights, and thus pleasant light and a pleasant atmosphere for the kitchen-cum-living room. It is as though the cool atmosphere has been left out; what we have here is a technological appliance that we can integrate without it making us feel uncomfortable.

BOSCH

Hersteller ***Manufacturer***	**Vertrieb** ***Distributor***	**Design** ***Designer***
Robert Bosch Hausgeräte GmbH Hochstraße 17 D-81669 München Tel. +49 89 45 90 01 Fax +49 89 45 90 23 47	Robert Bosch Hausgeräte GmbH Hochstraße 17 D-81669 München Tel. +49 89 45 90 01 Fax +49 89 45 90 23 47	Hans-Reinhart Janssen Werksdesign Robert-Bosch-Straße D-89537 Giengen Tel. +49 73 22 92 25 52

Wein-Lagerschrank „Bosch KSW 38920" *wine storage cabinet*

Mit diesem Wein-Lagerschrank können bis zu hundert 0,75 Liter Weinflaschen mit der richtigen Luftfeuchtigkeit und Temperatur, die normalerweise nur ein Naturkeller bieten kann, gelagert werden. Die Temperatur läßt sich stufenlos mit Kontrolleuchte von plus sechs bis plus vierzehn Grad regeln. Ein Aktivkohlefilter sorgt für dosierte Frischluftzufuhr und verhindert Fremdgerüche. Dank des eleganten, hochwertigen Designs paßt der Wein-Lagerschrank in jedes Küchen- und Wohnambiente. Er zeichnet sich durch ein silber-metallic lackiertes Gehäuse und eine gewölbte Front aus. Die doppelt isolierte, bombierte, getönte Glastür ist mit einem Alu-Rahmen mit alufarbenem Stangengriff eingefaßt. In einem schönen Kontrast zur sachlichen und kühlen Gestaltung stehen die sieben Holzablageböden im Innenraum, die sinnliche Wärme vermitteln und von denen sechs in der Höhe verstellt oder herausgenommen werden können. Der Wein-Lagerschrank ist mit einer Halogenbeleuchtung in der Blende ausgestattet. Außerdem besitzt er vorne höhenverstellbare Füße und hinten Rollen.

Up to one hundred 0.75 litre wine bottles can be stored in this wine storage cabinet at the right humidity and temperature, something that normally only a natural cellar can offer. Temperature can be set at any value between + 6°C and + 14°C, with a control lamp showing the temperature setting. An activated carbon filter ensures that the right amount of fresh air gets through, and forms a barrier against any foreign odours. Thanks to its elegant, high-quality design, the wine storage cabinet blends in with any kitchen or living environment. Its housing has a silver-metallic finish and a concave front panel. The doubly insulated, cambered and tinted glass door is set in an aluminium frame with an aluminium coloured handle. The seven wooden storage shelves inside the cabinet contrast nicely with the sober cool design of the exterior, giving a feeling of sensual warmth. Six of these shelves can be set at different heights, or taken out altogether. The wine storage cabinet is equipped with halogen lighting in the front panel. It has height-adjustable feet at the front and castors at the back.

Jury ***Judges panel***

Da der Weinkeller im Eigenheim fast die Ausnahme ist, ist dies eine adäquate Notlösung zur Lagerung von Wein. Versuch durch die angenehm gestalteten Holzböden noch eine Anmutung an frühere Weinvorratshaltung zu kreieren. Holz versus Aluminium. Das ist aber auch der Reiz.

As a wine cellar in your own home has almost become an exception, this is a reasonable alternative for storing wine. With its pleasantly designed wooden shelves it attempts to create an allusion to the old way of storing wine. Wood versus aluminium. But that is what makes it attractive.

SIEMENS
SIEMENS

Hersteller ***Manufacturer***	**Vertrieb** ***Distributor***	**Design** ***Designer***
Siemens Electrogeräte GmbH Hochstraße 17 D-81669 München Tel. +49 89 45 90 23 87 www.siemens-hausgeraete.de	Siemens Electrogeräte GmbH Hochstraße 17 D-81669 München Tel. +49 89 45 90 01 www.siemens-hausgeraete.de	Christoph Becke Werksdesign

Kühl-Gefrier-Kombination „KS 39 V 60" *fridge-freezer combination*

Schlichtes Design, zeitlose Erscheinung und edles Material charakterisieren die Reihe „steel-line", zu der auch diese Kühl-Gefrier-Kombination zählt. Diese fällt durch ihr hochwertiges Gehäuse aus geschliffenem Edelstahl auf. Schlichte senkrechte Türgriffe runden die kühle, sachliche und funktionale Formgebung der äußeren Erscheinung ab. Die Hochwertigkeit spiegelt sich auch in Technik und Ausstattung wider. So bringt die Null Grad-Frischkühltechnik professionelle Lagermethoden ins Haus. Im Inneren wurde sehr viel Wert auf gute Detailgestaltung gelegt. Es gibt beispielsweise ein Bottle-Rack aus geschliffenem Edelstahl. Die hochwertigen transluzenten Glasplatten sind in Edelstahl gefaßt und die Innentüreinteilung kann nach Bedarf variiert werden. Im Gefrierraum läßt sich der Abstellrost aus Edelstahl verstellen. Die Kühl-Gefrier-Kombination hat durch einen Innenventilator eine dynamische Kühlung.

The "steel-line" series, to which this fridge-freezer combination also belongs, is characterized by plain design, timeless appearance and high-quality materials. This appliance stands out by virtue of its high-quality housing of ground stainless steel. Plain vertical door handles round off the cool, sober and functional design of its exterior. High quality is also reflected in technology and equipment. Its superfast freezing system, for example, brings professional storage methods into the home. In the interior, a lot of importance and attention has been given to detail. There is, for example, a bottle rack of ground stainless steel. The high-quality, translucent glass shelves have a stainless steel surround, and the compartments on the inside of the fridge door can be varied as required. The stainless steel shelf in the freezer compartment can be set at different heights. With its internal fan, the fridge-freezer combination provides dynamic cooling.

Jury ***Judges panel***

Ein weiterer Klassiker im Kühlschrankbereich, wenig neue Details, dafür aber sehr überzeugend, auch sehr gute Innenaufteilung mit subtiler Materialkombination. Übersichtliche Innengestaltung, luftig, leicht, frisch. Große Gemüsefächer.

Another classic refrigerating appliance, not many new details, but what there is makes a lot of sense. The way the interior has been sub-divided is very good, with a subtle combination of materials. The interior design is clear, airy, light, fresh. Large salad crisper compartments.

Koch/Bunt
M1:Hemd 40°C
Pflegeleicht
M2:Mix.kalt
Fein
Memory3
Wolle
weitere
start
stop
serie IQ

Hersteller ***Manufacturer***	**Vertrieb** ***Distributor***	**Design** ***Designer***
Siemens Electrogeräte GmbH Hochstraße 17 D-81669 München Tel. +49 89 45 90 23 87 www.siemens-hausgeraete.de	Siemens Electrogeräte GmbH Hochstraße 17 D-81669 München Tel. +49 89 45 90 01 www.siemens-hausgeraete.de	Johannes Geyer, Gerd E. Wilsdorf Werksdesign

Waschmaschine „Serie IQ 1630" *washing machine*

Zum ersten Mal wird hier eine Waschmaschine über ein großzügiges Display bedient, das mit einfachen Worten in der gewünschten Landessprache durch die Programme führt. Die „Serie IQ" denkt über eine intelligente, lernfähige Bedienlogik mit. Neben einer Dosierempfehlung wird das Wäschegewicht automatisch angezeigt – so weiß man, wieviel Wäsche noch in die Maschine kann. Soll die Wäsche um 18 Uhr fertig sein, wird das einfach eingegeben. Das Waschprogramm startet dann so, daß es genau um 18 Uhr zu Ende ist. Neue, moderne Programme wie „Blusen/Hemden", „Microfaser" oder „Gardinen" sind zu den Basisprogrammen dazugekommen. Will man diese ändern, beispielsweise eine höhere Temperatur oder niedrigere Schleuderdrehzahl haben, stellt man sich einfach sein persönliches Wunschprogramm zusammen und speichert es mit der Memoryfunktion ab. Zu dem hohen Bedienkomfort trägt auch das großzügige Bullauge mit einem maximalen Öffnungswinkel von 180 Grad und der ergonomische Griff bei. Die Waschmaschine hat einen vierzig Zentimeter hohen Sockel mit einer Schublade, in der Waschmittel gelagert werden kann.

For the first time, a washing machine that is controlled by means of a generous display, which takes through the programmes s/he can select in simple words in his own language. With its intelligent, adaptive operating logic, "Serie IQ" follows its user's train of thought. Apart from recommending how much detergent to use, the weight of the load is automatically displayed – and so the user knows how much more washing will go into the machine. If the washing has to be done by 6 p.m., this is simply entered. The washing programme will then start so that it finishes at 6 p.m. on the dot. The basic programmes have been extended to include new, modern programmes such as "blouses/shirts", "microfibre" or "curtains". These can be changed (for example increasing the temperature or reducing spin-dryer speed) simply by putting your own personal programme together and using the memory function to save it. It is extremely convenient to use, not least because of the generous front-loading door with its ergonomic handle, which opens 180 degrees. The washing machine rests on a 40 centimetre high pedestal, containing a drawer in which detergent can be stored.

Jury ***Judges panel***

Eine Waschmaschine, die man im Wohnzimmer aufstellen kann, raus aus dem Keller, kann noch weiterentwickelt werden. Gute Lesbarkeit, große Funktionstasten, Minimierung auf die notwendigen Funktionen. Ein neues Detail, das integrierte Waschmittelfach, dadurch eine höhere Position des Stopflochs und damit eine einfachere Be- und Entladung der Waschmaschine.

At last a washing machine that can be put in the living room. Out of the basement with it, an idea that can be developed still further. Easy to read, large functional keys, reduced to necessary functions. The integrated drawer for detergent is a new detail that raises the height of the door and so makes it easier to load and unload the washing machine.

Hersteller ***Manufacturer***	**Vertrieb** ***Distributor***	**Design** ***Designer***
Duravit AG Werderstraße 36 D-78132 Hornberg Tel. +49 78 33 7 00 www.duravit.de	Duravit AG Werderstraße 36 D-78132 Hornberg Tel. +49 78 33 7 00 www.duravit.de	Frank Huster Eltviller Straße 18 D-65197 Wiesbaden Tel. +49 6 11 4 73 77 Fax +49 6 11 4 77 47

Behindertenwaschtisch „Architec H70" *disabled person's washstand*

Nicht zuletzt durch die Zunahme der älteren Bevölkerung gewinnt die barrierefreie Gestaltung öffentlicher Flächen, von Objektbauten aber auch von Wohnungen für Menschen, deren Bewegungsapparat aufgrund des Alters oder einer Behinderung eingeschränkt ist, immer mehr an Bedeutung. Eine große Rolle spielen hier speziell Bad und WC. Bei den meisten üblichen Sanitärprodukten aus dem Behindertenbereich wird aber noch häufig ohne Designanspruch und mit mangelnder Funktionalität gestaltet. Hiervon hebt sich der Waschtisch „Architec H70" ab. Er bietet neben einem eleganten, geräumigen, flachen Becken, das mit einem Rollstuhl unterfahrbar ist, auch seitliche Ablageflächen. Dank einer umlaufenden, integrierten und ergonomisch ausgeformten Greifreling kann der Rollstuhlfahrer sich ganz bequem mit einer Hand an den Waschtisch heranziehen. Durch das keramische Sieb, das optional mit Verteilerhaube abgedeckt werden kann, und den bewußten Verzicht auf einen Überlauf wird der Bakterienbildung vorgebeugt, was beispielsweise für Krankenhäuser von großer Bedeutung ist.

One major consequence of the greying of society is the ever increasing importance of the obstacle-free design of public spaces, of office buildings and of homes for people whose movement is restricted, whether because of age or because of a handicap. Bathrooms and toilets play an especially important role here. However, the vast majority of bathroom products for the disabled are frequently still made without any design appeal, and their functionality is poor. The "Architec H70" washstand is an exception to this rule. Apart from an elegant, large, shallow washbasin, which a wheelchair can fit under, it also has flat surfaces on each side. The bottom inside edge of the washstand has been ergonomically formed so that it can be gripped easily, allowing the wheelchair user to pull himself up to the washstand with one hand. The formation of bacteria is prevented by the ceramic strainer (which can be covered with a distributor hood if desired) and the deliberate avoidance of an overflow. This is especially important in hospitals, for example.

Jury ***Judges panel***

In Ergänzung des bestehenden Programms eine überzeugende flache Alternative für den Behindertenbereich. Und nicht nur für den. Es hat keineswegs die Anmutung von Behinderten-Sanitäranlage sondern ist eine einfache und charmante Beckenlösung, die in jedem Haushalt einen guten Platz finden kann.

As a complement to the existing range, a convincing shallow alternative for the disabled. But not only for them. It is by no means a typical piece of bathroom furniture for the handicapped, but is a simple, charming washbasin that would not be out of place in any household.

DORN
BRACHT

Hersteller *Manufacturer*	**Design** *Designer*
Dornbracht GmbH & Co. KG	Sieger Design
Kobbingsmühle 6	Schloß Harkotten
D-58640 Iserlohn	D-48336 Sassenberg
Tel. +49 23 71 43 30	Tel. +49 54 26 9 49 20
Fax +49 23 71 43 32 32	www.sieger-design.com

Elektronische Armatur „eMote" *electronic mixer tap*

Hand hin, Wasser an. Hand wieder hin, Wasser aus. Was bisher ausschließlich dem öffentlichen Bereich vorbehalten war, wird hier in die privaten Badezimmer gebracht: die berührungslose Funktionsweise. Das Wasser läuft und stoppt über einen Sensor, der auf Bewegung reagiert. Die elektronische Armatur „eMote" verbindet intelligente Technik mit hochwertigem Design, das Form und Funktion harmonisch vereinigt. Es gibt zwei Baugruppen, eine für den öffentlichen und eine für den privaten Sanitärbereich. Für öffentliche Anlagen wurde die Handhabung vereinfacht durch eine Automatik, die sich im Intervall von zwei Minuten ausschaltet. Für die privaten Nutzer gibt es eine on-off-Steuerung. Innovativ ist auch die Batteriebox, die üblicherweise in Plastik hergestellt wird, bei „eMote" aber aus Chrom besteht und damit der exklusiven Anmutung gerecht wird. Die Produktserie wird in zwei Designvarianten angeboten: eine Einlochmischvariante und eine Armaturen-Wandlösung.

Move your hand towards it, and the tap is switched on. Move your hand towards it again, and the tap is switched off. Something that was previously reserved for public washrooms has now made its entry into the private bathroom: non-touch operation. Water flows and stops via a sensor that reacts to movement. The "eMote" electronic tap fitting combines intelligent technology and high-quality design, which creates a harmony of form and function. Two assemblies are available, one for public and one for private bathrooms. For public use, the fitting has been simplified by adding an automatic function that switches the tap off after two minutes. For private use, there is an on-off control. A further innovative feature is the battery compartment. While this is usually made of plastic, it is made of chrome in "eMote", which does justice to its exclusive air. The product series is available in two design variants: a single-jet mixer variant, and a wall-mounted fitting.

Jury *Judges panel*

Zeigt die Kontinuität guter, bereits bekannter Lösungen in bestechender Form. Ermöglicht das Wassersparen auch im privaten Haushalt. Formal super gelöst, aber es gibt inzwischen so viele Versionen mit dieser Halbkreisbogenform. Stellt auch die Kontinuität unter Beweis, daß man in dieser Richtung immer noch qualitative Varianten entwickeln kann.

A fine example of the continuity of good, already known solutions. Allows water to be saved in the private household, too. Formally excellent solution, but we now have so many versions with this semi-circular form. Is also proof of continuity, that high-quality variants can still be developed in this direction.

98.4 FM
2
1
2
3
4
STATION
STATION
DISP MODE
FREQ SET
TIME SET
VOLUME
+
–
TIMER
hansgrohe

Hersteller ***Manufacturer***	**Vertrieb** ***Distributor***	**Design** ***Designer***
Hansgrohe AG	Hansgrohe AG	Phoenix Product Design
Auestraße 5–9	Auestraße 5–9	Kölner Straße 16
D-77761 Schiltach	D-77761 Schiltach	D-70376 Stuttgart
Tel. +49 78 36 51 14 93	Tel. +49 78 36 51 14 93	Tel. +49 7 11 95 59 76 11
www.hansgrohe.com	www.hansgrohe.com	Fax +49 7 11 55 93 92

Duschradio „Hansgrohe Oyster" *shower radio*

Das Bad entwickelt sich zunehmend vom reinen Zweckbad zu einem Wohlfühlraum. Ein Ort nicht nur um sich zu reinigen, sondern auch zu erfrischen und zu entspannen. Mit dem Duschradio kommt nun ein weiteres sinnliches Element ins Bad: Musik – um zu beruhigen, zu inspirieren oder zu beleben. Das Radio zeichnet sich durch ein kompaktes, stabiles Gehäuse aus, dessen Schale das Innenleben des Radios mit Senderspeicher vor Spritzwasser schützt. Die Technik im Inneren stammt von der Firma Grundig. Die angenehm runde Gestalt erinnert an maritime Formen, an die auch der Name „Oyster" (Auster) anknüpft. Dank der integrierten Öse läßt sich das Duschradio überall aufhängen. Man kann es aber auch frei aufrecht mit Hilfe eines Standfußes aufstellen oder durch eine Halterung direkt an die Wand montieren. Die externe Antenne wird für den optimalen Empfang mittels eines Saugknopfes an den Fliesen oder der Duschabtrennung fixiert. Strom bezieht es über drei Eineinhalb-Volt-Batterien. Es verfügt über einen Sendersuchlauf, vier Sender lassen sich zusätzlich abspeichern. Dank einer Timer-Funktion schaltet sich das Gerät nach einer vorgewählten Zeit automatisch ab.

Increasingly, the bathroom is developing away from its functional image and becoming a room in which to relax and feel good. Not just a place to cleanse oneself, but also a place of refreshment and relaxation. The shower radio now brings a further sensual element into the bathroom: music – to sooth, to inspire or to arouse. The radio has a compact, robust housing, whose outer shell keeps splashing water away from the internal workings of the radio, with its station-setting memory. The internal working parts are made by Grundig. Its pleasant round form is reminiscent of marine shapes, which are alluded to in its name "Oyster". The integrated eyelet means that the shower radio can be hung anywhere. It can also be stood up on its own with the help of a stand, or attached directly to the wall by means of a bracket. For optimum reception, the external aerial is fixed by a suction pad to tiles or shower curtain. It gets its power from three 1.5 volt batteries. It has a music scan function, and its memory has room for four stations. A timer function means that the radio switches itself off automatically after a pre-set time.

Jury ***Judges panel***

Ein knuffiges Objekt, übersichtlich, mit einer Anmutung von Frische und Wasser. Es macht das Radiohören im Bad geradezu sinnig. Bestechend schöne Materialkombination zwischen Silikon und einem Aluminiumersatz in Anlehnung an bestehende Duscharmaturen von der gleichen Firma. Gut und schön.

A cute object, easy to understand, alluding to freshness and water. It gives listening to the radio in the bathroom an almost sensual quality. Charming choice of materials, with silicon and aluminium surrogate quoting the existing shower attachments by the same company. Good and beautiful.

Hersteller ***Manufacturer***	**Vertrieb** ***Distributor***	**Design** ***Designer***
MAM Babyartikel GmbH	MAM Babyartikel GmbH	I. D. Pool; Ernst Beranek
Lorenz-Mandl-Gasse 50	Rudolf-Diesel-Straße 6	Singerstraße 6
A-1160 Wien	D-27383 Scheessel	A-1010 Wien
Tel. +43 1 49 14 10	Tel. +49 42 63 9 31 70	Tel. +43 1 5 12 58 41
www. mambaby.com	Fax +49 42 63 93 17 20	Fax +43 1 5 12 58 41 75

Lern-Zahnpflege Set *Beginner's dental care set*

Das Lern-Zahnpflege-Set ist für Babys ab sechs Monaten. Es wurde in Zusammenarbeit mit Kinderzahnärzten entwickelt und besteht aus zwei verschiedenen Zahnbürsten. Die extra lange „Lern-Zahnbürste" bietet genügend Platz, um sie gemeinsam mit dem Kind zu halten und ihm das richtige Zähneputzen beizubringen. Das kleinere Modell ist die erste eigene Zahnbürste. Damit kann das Kind, hat es erst das Zähneputzen erlernt, die Zähne selbst pflegen. Der weiche, abgerundete Zahnbürstenkopf paßt sich perfekt der kleinen Mundform des Babys an und schützt vor Verletzungen. Mit den kurzen, extra weichen, abgerundeten Borsten aus Polyamid läßt sich Zahnbelag entfernen, ohne den empfindlichen Zahnschmelz zu beschädigen. Die pinkfarbenen Borsten zeigen an, wieviel Zahnpasta verwendet werden sollte. Mit dem weichen Massagekopf läßt sich sanft Zunge und Zahnfleisch des Babys reinigen. Zudem kann das Baby darauf kauen, was Zahnungsschmerzen lindert. Der weiche Griff aus Polypropylen, das teilweise mit thermoplastischem Elastomer überzogen ist, ist ergonomisch geformt, leicht zu greifen und rutscht nicht. Ein Sicherheitsschild, das auf beide Modelle paßt, verhindert, daß die Zahnbürste zu tief in den Mund geschoben wird.

This beginner's dental care set is for babies from six months old. It was developed in collaboration with children's dentists, and comprises two different toothbrushes. The extra-long "learner's toothbrush" has space enough for it to be held together with the child, and to teach it how to clean its teeth properly. The smaller model is the first proper toothbrush. Once the child has learnt how to clean its teeth, it can use this on its own. The soft-rounded head of the toothbrush is perfectly adapted to baby's small mouth, and protects it against injury. The short, extra-soft, rounded polyamide bristles allow film to be removed without damaging the sensitive enamel. The pink bristles show how much toothpaste should be used. The soft massage head allows gentle cleaning of the baby's tongue and gums. And baby can chew on it to alleviate teething trouble. The soft polypropylene grip, part of which is covered with thermoplastic elastomer, is ergonomically shaped, easy to grasp and non-slip. A safety plate that fits on both models prevents the toothbrush being pushed too deep into baby's mouth.

Jury ***Judges panel***

Eine Baby-Zahnbürste mit einem zusätzlichen Gummikopf zur Zahnfleischmassage und verschiedenen anderen interessanten Details, wie dieser „Handstopper", damit sie nicht zu weit in den Mund gesteckt wird. Ein Produkt, das Bedürfnisse von Menschen ernst nimmt, die sie noch nicht selbst artikulieren können, also Kleinkinder. Kinderfreundliche Ergonomie, muß das sein?

A baby's toothbrush with an additional rubber head to massage the gums and various other interesting details, such as this "hand-stopper" preventing it being pushed too far into the mouth. A product that takes seriously the needs of people who cannot yet articulate themselves, i.e. infants. Child-friendly ergonomics, do we really need this?

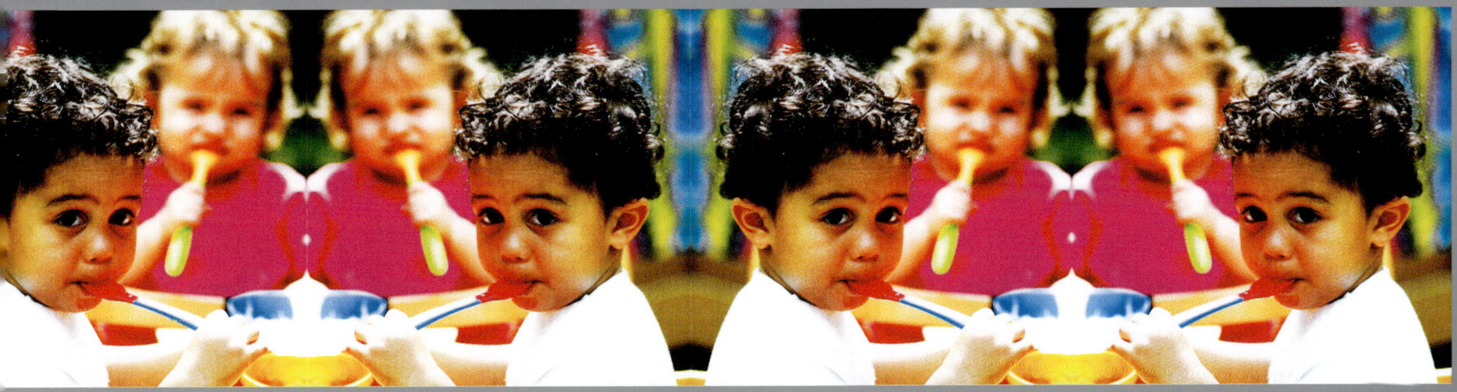

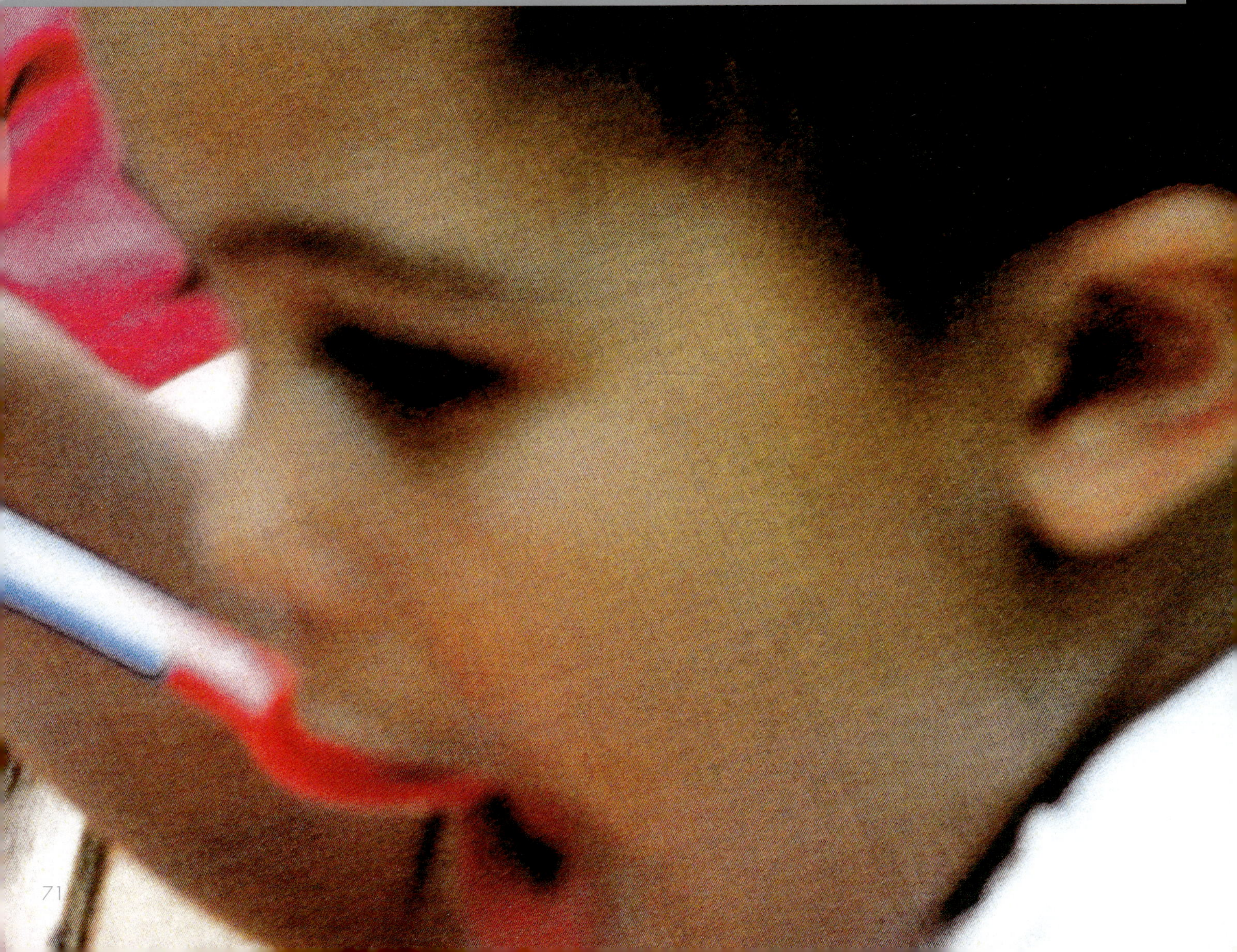

Focus in Gold

Hersteller *Manufacturer*	Vertrieb *Distributor*	Design *Designer*
BabyBjörn AB Box 595 S-18236 Danderyd Tel. +46 8 54 49 68 00 Fax +46 8 54 49 68 29	Sales Plus BabyBjörn Deutschland Wetterkreuz 5 D-91058 Erlangen Tel. +49 1 80 55 16 00	Ergonomi Design Gruppen Håkan Bergkvist Box 14021 S-16141 Bromma Tel./Fax +46 8 26 22 41

Fütterlöffel „BabyBjörn®" feeding spoon

Der Fütterlöffel für Babys ist gut durchdacht und verbindet gutes Design mit hoher Funktionalität und Sicherheit. So ist das Löffelblatt weich und geschmeidig, was dem noch empfindlichen Zahnfleisch von Babys gerecht wird. Mit dem Löffel läßt sich auch ganz leicht Brei, der neben dem Mund des Babys gelandet ist, wieder abstreifen. Dank dem biegsamen Löffelblatt kann man das Essen leicht vom Boden eines Tellers oder eines Gläschens aufnehmen. Das hintere Ende des Löffelblattes ist erhöht, was verhindert, daß das Essen vom Löffel fällt, wenn dieser nach hinten kippt. Durch die Form des Löffels läßt sich auch in hohen Babygläschen der Boden erreichen. Zudem liegt der Löffel angenehm in der Hand der fütternden Person. Da das Löffelblatt dünn und doch tief ist, kann man den Löffel nicht nur für das Baby sondern sehr lange weiter nutzen. Das recyclebare Material des Löffels ist PVC-frei.

This feeding spoon for babies has been well thought through, and combines good design with a high level of functionality and safety. The bowl of the spoon is soft and supple, and so is suitable for babies' still sensitive gums. Any broth that has not landed in the baby's mouth can also easily be wiped away again with the spoon. Thanks to the spoon's flexible bowl, food can easily be scraped up from the bottom of a dish or jar. The rear end of the spoon bowl is raised, and so prevents food dropping from the spoon if it is tipped backwards. The spoon's form means that even the bottom of tall jars of baby food can be reached. Moreover, the spoon fits comfortably into the hand of the person feeding the baby. As the bowl of the spoon is thin yet deep, the spoon can be used for a long time, and not only for babies. The recyclable material of which the spoon is made is PVC-free.

Jury ***Judges panel***

Hier hat jemand über die Bedürfnisse derjenigen nachgedacht, die sich noch nicht selbst artikulieren können. Wunderschöne Verarbeitung, etwas Althergebrachtes neu überdacht und verbessert. Äußerst reduzierte Kombination zweier extremer Kunststoffmaterialien zeigt auch die Technologiespannbreite im Kunststoffbereich.

Here, someone has thought about the needs of those who cannot yet express what they want. Beautiful finish, a reconsideration and improvement of something that has been around a long time. Extremely reduced combination of two extreme plastic materials is also evidence of the range of technology in the area of plastics.

AUERHAHN

Hersteller ***Manufacturer***	**Vertrieb** ***Distributor***	**Design** ***Designer***
Auerhahn Bestecke GmbH Postfach 1161 D-72206 Altensteig Tel. +49 74 53 94 68 34 www.auerhahn-bestecke.de	Auerhahn Bestecke GmbH Postfach 1161 D-72206 Altensteig Tel. +49 74 53 94 68 34 www.auerhahn-bestecke.de	Peter Bäurle Wiesentalstraße 43/1 D-73312 Geislingen/Eybach Tel. +49 73 31 6 44 30

Menübesteck „Silk" *cutlery set*

„Silk" – das Besteck mit Oberflächen wie Samt und Seide. Erreicht wird diese Optik durch die feine Satin-Skin Mattierung, die dem Besteck zusammen mit den wohlgeformten weichen fließenden Linien etwas Sinnliches gibt. Puristisches, klares Design und schlichte, zeitlose Gestaltung verbindet sich hier mit einer organischen Formensprache. Der sanfte charakteristische Schwung zieht sich durch das Besteck bis in die gerundeten Spitzen. „Silk" ist aus hochwertigem Edelstahl, rostfrei und spülmaschinenfest.

"Silk" – cutlery with a surface like velvet and silk. This appearance is brought about with the help of its fine satin-skin depolishing which, together with the elegant form of its soft, flowing lines, gives the cutlery a sensual character. Purist, clear design and an unobtrusive, timeless form combine here with a biomorphic formal language. The soft characteristic sweep is present right through the cutlery set, right down to its rounded points. "Silk" is made of high-quality stainless steel, is corrosion-resistant and dishwasher-safe.

Jury ***Judges panel***

Eine Bestecklösung mit Eleganz und Grazie, auch die Oberflächenqualität stimmt, gebürsteter Edelstahl. Es liegt gut in der Hand, die Gewichte sind gut austariert. Immer wieder bestechend, mit welch einfachen Mitteln man eigentlich althergebrachte klassische Ikonen differenziert.

Cutlery with elegance and grace, the surface quality is right too, brushed stainless steel. It sits comfortably in the hand, its weight is well balanced. It is always fascinating to see how simply a new twist can be given to traditional, classical icons.

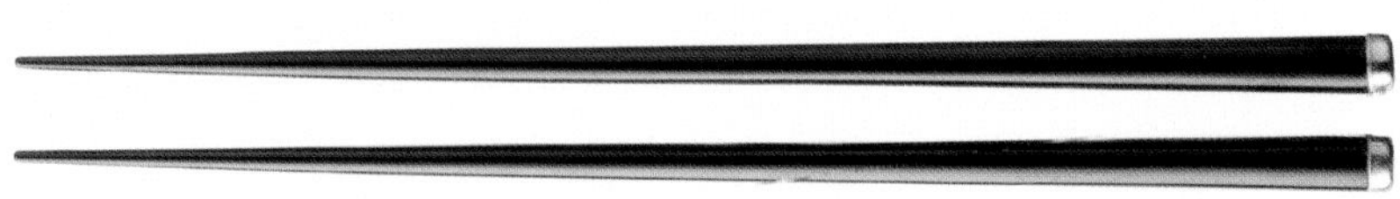

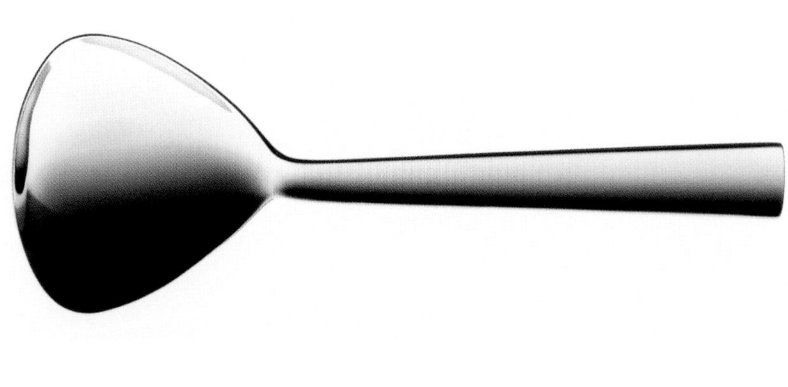

Hersteller *Manufacturer*	Vertrieb *Distributor*	Design *Designer*
WMF Württembergische Metallwarenfabrik AG Eberhardstraße D-73312 Geislingen/Steige Tel. +49 73 31 2 51 www.wmf.de	WMF Württembergische Metallwarenfabrik AG Eberhardstraße D-73312 Geislingen/Steige Tel. +49 73 31 2 51 www.wmf.de	Ole Palsby Strandvejen 203 DK-2900 Hellerup Tel. +45 39 62 00 09 Fax +45 39 62 53 18

Asia-Besteckset „CombiNation" *Asian cutlery set*

Die Küche des Mittelmeerraumes ist mittlerweile zum festen Bestandteil unserer Eßgewohnheiten geworden, und auch asiatische Gerichte nehmen immer mehr Einzug in unsere Küche. Die Speisen sind vielfältiger geworden, nicht jedoch die Werkzeuge. Dem trägt „CombiNation" – das Besteck der 1000 Möglichkeiten – Rechnung. Aus 54 Besteckteilen kann sich jeder sein individuelles Eßwerkzeug zusammenstellen. Ein Teil davon ist das von der Jury ausgezeichnete Asia-Besteckset. Es besteht aus einem Servierlöffel mit einer dreieckigen und Reislöffeln mit tiefer ovaler Laffe. Dazu gehören Chopsticks und Schälchen. Bei den Gabeln gibt es neben der herkömmlichen Version mit vier Zinken eine fünfzinkige etwa für Reis oder Erbsen. Eine dreizinkige Gabel eignet sich zum Aufspießen zerkleinerter Nahrungsmittel, etwa Fleisch. Die breite Auswahl zieht sich ebenso bei Löffeln und Messern durch. Jedes Teil paßt zum anderen. „CombiNation" befriedigt damit die Lust am Kombinieren. Das Besteckset ist klar, sachlich, schlicht, zeitgemäß und funktional gestaltet. Die leichte Wölbung und abgerundete Kanten machen es weich und lassen es angenehm in der Hand liegen.

Mediterranean cuisine has long since become a well established part of our diet, and Asian dishes are also increasingly finding their way into our kitchens. Food has become more varied, but not the tools to eat it with. "CombiNation" – the cutlery of 1000 possibilities – is the answer to this. Anyone can put together their individual cutlery combination from the 54 cutlery parts. The Asian cutlery set commended here is part of this. It comprises a serving spoon with a triangular bowl, and rice spoons with a deep oval bowl. It is complemented by chopsticks and small bowls. For the forks, there is the traditional four-tined model and a model with five tines for food such as rice or peas. A three-tined fork is useful for picking up small pieces of food, such as meat. This wide range is also to be found in the spoons and knives. Each part goes with the other. "CombiNation" therefore satisfies the wish to combine. The design of the cutlery set is sober, unostentatious, modern and functional. Its slightly convex form and rounded edges give it soft contours and make it sit comfortably in the hand.

Jury ***Judges panel***

Interessanter Beitrag zum Thema Asien und Globalisierung, weitgehend drei Materialien zusammen verknüpft, die eine erstaunliche Harmonie abgeben, im Detail sorgfältig gestaltet.

Interesting contribution to the topics of Asia and globalization, for the most part three materials are combined surprisingly harmoniously, careful attention to detail in the design.

Hersteller ***Manufacturer***	**Vertrieb** ***Distributor***	**Design** ***Designer***
M. H. Wilkens & Söhne GmbH An der Silberpräge D-28305 Bremen Tel. +49 4 21 4 10 32 73 Fax +49 4 21 4 10 31 30	M. H. Wilkens & Söhne GmbH An der Silberpräge D-28305 Bremen Tel. +49 4 21 4 10 32 73 Fax +49 4 21 4 10 31 30	Claudia Christl Werksdesign

Kinderbesteck „Baba" *child's cutlery*

„Baba" ist das erste Kinderbesteck in Edelstahl der Firma. Zum Besteck für Kinder wird es sowohl durch die zurückgenommene Größe als auch durch die Proportionen. Die Vorderteile, also Löffellaffe und Klinge, sind durch ihre Größe betont, deuten mit ihrer Form auf die verschiedenen Funktionen hin und unterstützen diese.
„Baba" besitzt ein eigenständiges Design, das durch asymmetrische Konturenverläufe auffällt. Die Formgebung ist weich, fließend und gedrungen, was nur durch den schmalen Hals zwischen Vorderteil und Stiel durchbrochen wird. Dieser schmale Hals macht das unerwünschte Wandern der Speisen in Richtung Hand schwer. Das Besteck erfüllt ergonomische Grundbedürfnisse und läßt sich angenehm von Kinderhänden greifen. Es gibt keine Ecken und Kanten. Die asymmetrische Stiele laden sowohl Rechts- als auch Linkshänder zum Anfassen und Greifen ein.

"Baba" is the company's first stainless steel child's cutlery set. Both its reduced size and proportions mark it out as child's cutlery. The front parts, i.e. the spoon bowl and blade, are emphasized in terms of size, and their shape alludes to and supports their various functions.
"Baba" has an unusual design, with striking asymmetrical contours. Its soft, fluid and squat form is interrupted solely by the narrow neck between the front section and the handle. This narrow neck makes it difficult for any food to move up towards the hand. The cutlery satisfies fundamental ergonomic needs and is pleasant for a child's hand to grip. There are no corners or sharp edges. The asymmetric handles are an invitation to both left-handed and right-handed children to touch and grip.

Jury ***Judges panel***

Kindgerecht, Material hochwertig, gute Proportion, gute Balance, kleine Handschmeichler.

Child-friendly, high-quality material, well-proportioned, good balance, a joy to hold.

Pure 301 Steel
CHROMA
DESIGN BY
F.A.PORSCHE

Hersteller *Manufacturer*	**Vertrieb** *Distributor*	**Design** *Designer*
Chroma Cutlery Inc. 184 Oakland Ave. Suite B Spartanburg SC 29302, USA Tel. +1 8 88 2 50 88 09 www.happy-cooking.com	CRR Christian Romanowski Rockstedt e. K. Haupststraße 16 D-27404 Rockstedt Tel. +49 18 03 59 59 59 www.kochmesser.de	Ferdinand A. Porsche Porsche Design GmbH Flugplatzstraße 29 A-5700 Zell am See Tel. +43 65 42 57 22 70 info@porsche-design.at

Küchenmesser „Type 301" *kitchen knife*

Der Entwurf der Küchenmesserserie basiert auf ergonomischen Studien. „Messergreifen" ist im vorderen Bereich der Hand, das heißt Daumen gegen Zeigefinger, vertikal orientiert, dagegen im hinteren Bereich, also der kleine und der Ringfinger gegen den inneren Handballen, eher horizontal. Deshalb wollte man Schneide und Griff optisch nicht voneinander trennen, sondern entschied sich, die vertikale Fläche der Klinge in den Griff einlaufen zu lassen. Um jedoch aus Sicherheitsgründen anzuzeigen, hier endet der Griff, und hier beginnt das Messer, wurde als optisches und haptisches Signal die „Perle" eingebaut. Je nach Größenverhältnis von Klinge und Griff können einige Messer der Serie mit der Schneide nach oben auf dem Tisch stehen statt wie herkömmliche Messer liegen. Das hat den Vorteil, daß sie den Tisch nicht beschmutzen. Der Griff wird aus hochwertigem Edelstahl und die Klinge aus feinem japanischen Pure 301 Steel hergestellt.

Ergonomic studies were used as the basis for the design of this kitchen knife series. "Gripping a knife" comprises the vertical relationship between thumb and index finger in the front part of the hand and, in the rear part of the hand, the more horizontal relationship between little finger/ring finger and the palm of the hand. This was the reason why it was decided not to separate the cutting edge and the handle optically, but to allow the vertical surface of the cutting edge to continue on into the handle. For safety reasons, however, the "pearl" was integrated as an optical and tactile signal to show where the handle ends and the knife begins. Depending on the size relationship between blade and handle, some knives in the series can be left on the table with the cutting edge uppermost, instead of lying flat like conventional knives. The advantage is that they do not make the table dirty. The handle is made of high-quality stainless steel, and the blade of fine Japanese Pure 301 Steel.

Jury ***Judges panel***

Das komplette Messer wie die Klinge durchgängig aus Stahlmaterial. Auch die Interpretation der Ergonomie des Griffes ist absolut überzeugend. Dadurch wird auch eine neue ästhetische Anmutung erreicht.

The complete knife made solely of steel material, like the blade. The way the ergonomics of the handle has been interpreted is absolutely convincing, too. The result is a new aesthetic gracefulness.

Hersteller ***Manufacturer***	**Vertrieb** ***Distributor***	**Design** ***Designer***
C. Hugo Pott GmbH & Co. KG Ritterstraße 28 D-42659 Solingen Tel. +49 2 12 4 30 56 www.pott-bestecke.de	C. Hugo Pott GmbH & Co. KG Ritterstraße 28 D-42659 Solingen Tel. +49 2 12 4 30 56 www.pott-bestecke.de	Ralph Krämer Leberstraße 8 D-66121 Saarbrücken Tel. +49 6 81 81 79 89 Fax +49 6 81 81 68 53

Serviergabel „Pastina" *serving fork*

„Pastina" ist das Ergebnis der Überlegung, daß, was das Beste ist, um Pasta zu essen, auch das Beste sei, um Pasta zu servieren. So kam es zu der ersten Serviergabel: ein überdimensioniertes Modell. Stiel und Funktionsteil drehen sich in einer Achslinie rund, das bedeutet, die Zinken wickeln die Spaghetti auf ohne zu „eiern". Die Zinken sind vorne rund geschliffen, wodurch sie die Pasta nicht aufspießen. Der Abstand zwischen den Zinken ist so gewählt, daß sich die Spaghetti beim Aufnehmen aufrollen, aber beim Servieren nicht zwischen den Zinken hängenbleiben, sondern glatt wieder abrutschen. Das Vorderteil fließt sanft in den Stiel über, es gibt keinen Absatz, an dem die Nudeln hängenbleiben können. Stiel und Vorderteil sind konkav geschwungen. Nimmt man die aufgerollte Pasta aus einer Schüssel, zeigen die Zinkenspitzen nach oben. So rutschen die Spaghetti nicht ungewollt ab. Beim Servieren dagegen zeigen sie sinnvollerweise nach unten. Im Gegensatz zu einem runden Stab, bei dem der Ansatzpunkt fehlt, läßt sich der Stiel dank seiner richtig gewählten Breite gut drehen. Das schmal zulaufende, aber gerundete Griffende hilft den Stiel beim Drehen gut zu führen.

"Pastina" arose from the idea that what is best when eating pasta must also be best when serving it. The result was the first ever spaghetti-serving fork: an oversize model. The handle and "fork" turn on one axis, which means that the tines pick up the spaghetti without wobbling. The tines are rounded at their ends, so that they do not spear the pasta. The tines have been spaced so that the spaghetti is wound when the serving fork is turned but does not stick between the tines when serving, sliding smoothly from the fork again. The transition between front section and handle is smooth, there are no projecting parts on which the noodles could get stuck. The handle and front section follow a concave curve. When taking the wound pasta from a bowl, the tines point upwards, and so the spaghetti cannot accidentally slide off. When serving, by contrast, the tines point downwards. Unlike a round rod which has no place to grip, the handle has just the right width, and can be turned easily. The narrowly tapered but rounded end of the handle helps guide the handle when it is being turned.

Jury ***Judges panel***

Schlicht und einfach. Für die Anwendung eine atypische Form, die aber durchaus der Aufgabe gerecht wird. Das nicht alltägliche Werkzeug verleitet zum Schmunzeln. Überraschende Besteckikone, immer seltener bei soviel Besteckproduktion. Überraschend dadurch, daß durch Größenverschiebungen und andere Gestaltung der Zinken unheimlich viel Attraktivität entsteht.

Plain and simple. An atypical form for this use, but one which certainly fulfils the task well. This out-of-the-ordinary implement gives cause to smile. Surprising cutlery icons: they are increasingly rare with so much cutlery being produced. Surprising because shifts in size and other formal work on the tines result in something incredibly attractive.

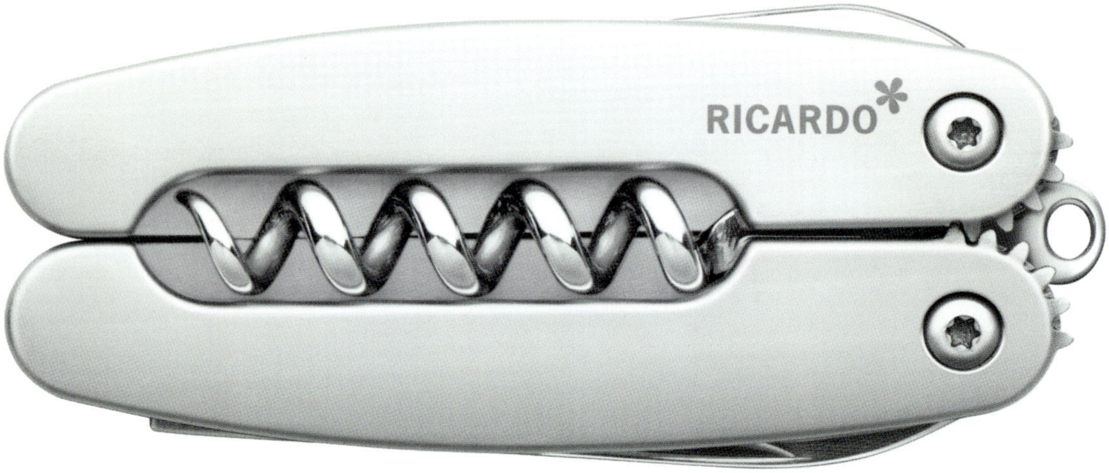
RICARDO

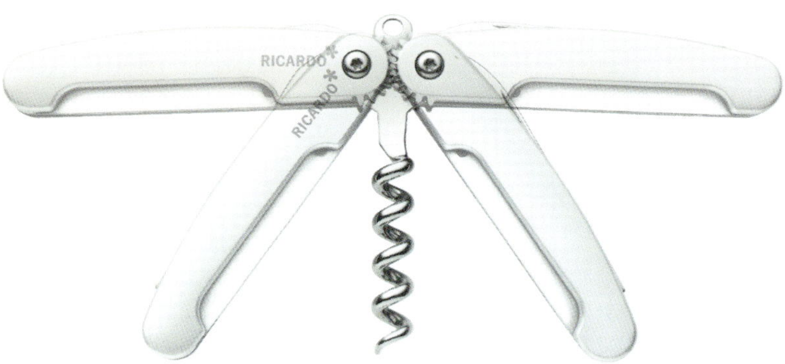
RICARDO
RICARDO

BOUTEILLE
DANS NOS
RICARDO

Hersteller ***Manufacturer***	**Vertrieb** ***Distributor***	**Design** ***Designer***
Gebr. Richartz + Söhne GmbH Merscheider Straße 94 D-42699 Solingen Tel. +49 2 12 23 23 10 www.richartz.com	Gebr. Richartz + Söhne GmbH Merscheider Straße 94 D-42699 Solingen Tel. +49 2 12 23 23 10 www.richartz.com	Vistapark GmbH Villa Media Viehhofstraße 119 D-42217 Wuppertal Tel. +49 2 02 2 42 75 00

Taschenmesser „Vintura" *pocket knife*

„Vintura" – das Weintool für den Schlüsselbund – sticht bereits durch seine geringen Maße hervor: es hat gerade mal eine Länge von sechs und eine Breite von nicht ganz zweieinhalb Zentimetern und ist nur ganze achtzig Gramm schwer. Trotzdem bietet es sehr viel: nicht nur Klinge, Minischere, Nagelfeile mit Schraubendreher sondern auch einen Korkenzieher für unterwegs, der sich einfach und bequem ausklappen läßt. Alle Teile sind aus rostfreiem Stahl hergestellt. Die Schalen sind aus Zinkdruckguß matt verchromt.

The small dimensions of "Vintura" alone are enough to make you sit up and take notice: it is just six centimetres long and not even two-and-a-half centimetres wide, and weighs a mere eighty grams. Nevertheless, this wine tool for your key ring has a lot to offer: not only blade, miniature scissors, or nail file with screwdriver, but also a corkscrew for the road, which is simply and conveniently opened out. All parts are made of stainless steel. The shells are made of matt chromium-plated die cast zinc.

Jury ***Judges panel***

Der Zigarrenschneider für Weinliebhaber, handlich, klein, allzeit bereit.
Schönes Taschenspielzeug.

The cigar-cutter for wine lovers, handy, small, always at hand. Nice pocket toy.

Hersteller ***Manufacturer***

Alfred Scholze GmbH
Gürtlerstraße 6
D-87600 Kaufbeuren,
Neugablonz
Tel. +49 83 41 9 77 00

Vertrieb ***Distributor***

Berendsohn AG
Alte Sülldorfer Landstraße 400
D-22559 Hamburg
Tel. +49 40 81 90 90
www.berendsohn.com

Design ***Designer***

Alfred Scholze
Werksdesign

Champagnerverschluß „Focus" *champagne bottle stopper*

„Focus" verschließt Champagnerflaschen, damit das edle Getränk auch nach dem ersten Öffnen nichts von seinem prikkelnden Genuß verliert. Die besondere Verriegelungsmechanik verschließt den Champagner druckdicht und hält ihn so frisch. Dezente zurückhaltende Eleganz und auf die Funktion reduziertes Design zeichnen den Champagnerverschluß aus und runden eine formvollendete Tischkultur ab. Neben Champagnerflaschen lassen sich mit „Focus" auch Sektflaschen verschließen.

"Focus" seals champagne bottles so that this precious drink loses none of its sparkling quality after being opened. The special sealing mechanism pressure seals the champagne bottle, keeping the champagne fresh. The stopper's appeal lies in its modest, unostentatious elegance and in its purely functional design; it provides a perfect finishing touch to any table setting. "Focus" can be used to seal sparkling wine bottles as well as champagne bottles.

Jury ***Judges panel***

Endlich ein salonfähiger Champagnerverschluß, der seinem Unterteil gerecht wird.

At last, a presentable champagne stopper that reflects the class of the bottle's contents.

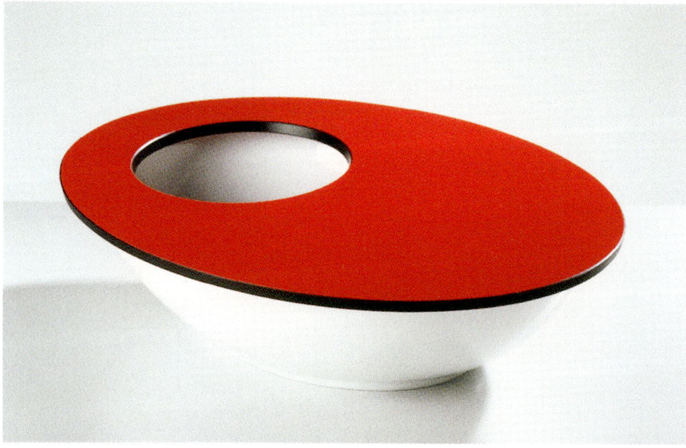

Hersteller ***Manufacturer***

Rosenthal AG
Wittelsbacherstraße 43
D-95100 Selb
Tel. +49 92 87 7 23 44
www.rosenthal.de

Design ***Designer***

David Queensbury + Martin Hunt
63 Penfold Street
London NW8 8PQ

Porzellan „Loft Complete" *china*

Internationale Food-Trends wie Pasta, Sushi oder Fusion Cooking und moderne veränderte Lebensgewohnheiten haben neue Standards gesetzt. Erlaubt ist, was Spaß macht. „Loft" ist eine Antwort darauf. Das moderne und zeitgemäße Short Set ist multifunktional und international. Unterschiedlich große und verschieden geformte Geschirrteile fügen sich in ein Gesamtkonzept, das freies Kombinieren erlaubt: runde Elemente etwa für Pasta, Salat, Müsli oder Suppen. Wer mag, kann auf ovalen Platten die Fischsuppe oder die Sushi-Häppchen servieren. Die unterschiedlich großen quadratischen Schalen lassen sich vor allem bei Vorspeisen, Dips und Knabbereien effektvoll einsetzen. Eckige Schüsseln und Glasschalen für Salat oder Vorspeisen funktionieren zum Beispiel auch als Deckel für eine unkomplizierte Aufbewahrung. Vielseitig einsetzbar, beispielsweise als Tablett, ist ein Multifunktionsbrett mit einem Loch, in das das eine oder andere Geschirrteil hineinpaßt. Oder als Schneidebrett: hier kann das geschnittene Gemüse direkt in die darunterstehende Schüssel geschoben werden. „Loft" zeichnet sich durch eine charakteristische Rillenstruktur aus. Es ist in weiß oder mit verschiedenen Dekoren zu haben.

International food trends such as pasta, sushi or fusion cooking and modern changes in living habits have set new standards. Anything is permitted, as long as it is fun. "Loft" is one response to this. The modern, contemporary "short set" is multifunctional and international. Crockery items of various shapes and sizes blend together in one overall concept that encourages all kinds of combinations, such as the round elements for pasta, salad, muesli or soups. Depending on mood, the oval plates can be used to serve fish soup or sushi titbits. The different sized square bowls can be used to good effect for hors d'oeuvres, dips and nibbles. Angular dishes and glass bowls for salad or hors d'oeuvres can also be used as lids, for example, making it easy to keep food. One item with a multitude of uses, for example as a tray, is a multifunctional board with a hole into which some of the chinaware items fit. Or as a cutting board: chopped vegetables can be pushed through the hole into a bowl. "Loft" has a characteristic grooved texture. It is available in white or in various patterns.

Jury ***Judges panel***

Oldtimer neu und intelligent aufgelegt. Klassische, schöne Formen, bei denen die Proportionen elegant und harmonisch gelöst sind. Der Vorteil an der Terrinenlösung ist zum Beispiel, daß es ohne Deckel eine Schüssel ist. Die Einsatzmöglichkeiten sind nicht zu spezifisch festgelegt.

An old friend in a new, intelligent version. Classical, attractive forms, which have been proportioned elegantly and harmoniously. For example, the way the tureen has been designed means that it can be a simple bowl if its lid is taken away. Uses have not been defined too strictly.

Bits

Hersteller ***Manufacturer***

Rosenthal AG
Wittelsbacherstraße 43
D-95100 Selb
Tel. +49 92 87 7 23 44
www.rosenthal.de

Design ***Designer***

Vogt & Weizenegger
Mulackstraße 15
D-10119 Berlin
Tel. +49 30 6 14 21 66
Fax +49 30 6 14 22 87

Porzellan-Sets „Units" *chinaware sets*

„Units" steht für moderne Lebensgewohnheiten, ästhetisch gelungenes Design, gute Funktionalität und ein auf einige wesentliche, multifunktional einsetzbare Teile beschränktes Geschirr: nur sechs stapelbare Einzelteile in ganz reduzierter Formensprache, die alles mitmachen vom Frühstück bis zum späten Abend. So gibt es drei verschiedene Becher für Espresso, Tee oder Milchkaffee. Diese schlichten Gefäße sind auch ideal für Snacks, Dips, Joghurt oder Müsli. Zu dem Porzellan-Set gehören noch eine kleine Schüssel für Suppen, Pasta und Salat, sowie zwei unterschiedlich große Teller. Dazu gibt es einen Glasbecher, der formal perfekt mit den Porzellanteilen harmoniert. Abgerundet wird die Kollektion durch originelle Ergänzungsteile aus hochwertigem Kunststoff, wie eine rote Abdeckung mit Strohhalm zu dem großen Becher, um ein heißes Getränk mit auf den Weg zu nehmen, oder der rote Griff, den man an den mittleren Becher und das Glas anklicken kann. Neben der weißen Variante gibt es drei Dekore: „Colours" mit den Farben dottergelb, rot und blau, „Smiley" mit lachenden Blümchen der siebziger Jahre oder „Moiré" mit Farbmix und Stilelementen der sechziger Jahre.

"Units" stands for modern living habits, aesthetically pleasing design, good functionality and china that has been reduced to a few essential items with a multitude of uses: just six stackable individual pieces of crockery, formally very simple, are enough for every need from breakfast to the late evening. There are three different mugs, for example: for espresso, tea or café latte. These unassuming vessels are also ideal for snacks, dips, yoghurt or muesli. The set is completed by a small bowl for soup, pasta or salad and two plain plates in two different sizes. There is also a glass beaker whose form harmonizes perfectly with the chinaware. The collection is rounded off by original accessories of high-quality plastic, such as a red lid with integrated straw for the large mug, allowing a hot drink to be drunk while on the move, or the red handle that can be clicked onto the medium-sized mug or the glass. Apart from white, the china is available in three patterns: "Colours", in yolk yellow, red and blue, "Smiley" with 1970s-style smiling flowers, or "Moiré" with 1960s-style colour combinations and style elements.

Jury ***Judges panel***

Weiterer Beitrag zum Thema „Neue Eßgewohnheiten" und weg mit Omas Tafelservice. Kleines Geschirrprogramm für jüngere Benutzer, wenige Teile erfüllen sehr viele Funktionen. Ein Oldtimer revitalisiert, intelligente Vorteile: Uralt-Stapelmethode der bekannten Hotelgeschirre in den privaten Haushalt gebracht, Mix des Materials Glas, Kunststoff, Porzellan, unter Nutzung neuester Technologien.

An important contribution to the "new eating habits" theme: away with granny's tea set. A small range of china for younger users, a few parts fulfil many functions. An old friend with a fresh image, intelligent advantages: the stacking method that hotels have been using for years brought into the private household, a mix of glass, plastic, china, using modern technology.

Hersteller ***Manufacturer***	**Vertrieb** ***Distributor***	**Design** ***Designer***
Eva Denmark A/S Hoejnaesvej 59 DK-2610 Roedovre Tel. +45 36 73 20 60 www.evadenmark.com	Eva Deutschland Pinneberger Chaussee 99 D-25436 Moorrege Tel. +49 41 22 8 18 62 Fax +49 41 22 8 92 22	Claus Jensen & Henrik Hoelbæk Tools Design Rentemestervej 23 A DK-2400 Kopenhagen NV Tel. +45 38 19 41 14

Thermosflasche „Eva Solo" *vacuum flask*

Die Thermosflasche verbindet elegantes Design mit der neuesten Technologie. Sie hat keinen Glaseinsatz mehr, der durch einen Kälte- oder Wärmeschock oder wenn man sie fallen läßt, in tausend Stücke zerbrechen kann. Die Thermosflasche dagegen ist bruchsicher, da sie innen wie außen aus rostfreiem Edelstahl besteht. Dank dem luftdichten Raum zwischen Flasche und Einsatz kann sie sehr viel länger Wärme oder Kälte der Getränke halten. Außerdem besitzt sie einen gießfreundlichen, tropffreien Rand. Dabei wird der letzte Tropfen in einer Art „Sicherheitsnetz" aufgefangen. Dieses patentierte System macht die Firma möglicherweise zum einzigen Hersteller von tropfsicheren Thermoskannen weltweit. Die Thermosflasche erinnert an die Form einer südländischen Karaffe für Wasser und Wein. Sie sticht durch ihre moderne zeitgemäße elegante Anmutung und die Materialmischung Edelstahl mit schwarzem hitzebeständigem Kunststoff hervor. Daneben gibt es auch eine Version ganz in schwarz. Die besondere Konstruktion der Flasche führt dazu, daß der Rauminhalt, nämlich 1,1 Liter, im Verhältnis zu ihrer Größe sehr groß ist. Dank des niederen Schwerpunktes ist sie äußerst standfest. In der gleichen Form gibt es auch eine Thermoskanne mit Griff.

This vacuum flask combines elegant design with state-of-the-art technology. It no longer has any glass liner that can break into a thousand pieces as a result of a temperature shock or of being dropped. Instead, this vacuum flask is unbreakable, as it is made inside and out of stainless steel. Thanks to the vacuum between the flask and the liner, it can keep drinks hot or cold much longer. In addition, its lip is designed so that it pours well without dripping. The last drop is caught in a kind of "safety net". This patented system may mean that the company is the only manufacturer of non-drip vacuum flasks in the whole world. The shape of the vacuum flask is reminiscent of a Mediterranean carafe for water or wine. It catches the eye with its modern elegance and grace, and with its mixture of stainless steel and black, heat-resistant plastic. There is also an all-black version. The special design of the flask means that its capacity (1.1 litres) is very big in comparison with its size. Thanks to its low centre of gravity, it is extremely stable. There is also a vacuum flask of the same shape with a handle.

Jury ***Judges panel***

Eine Urform des Mittelmeerraumes für Getränke, modern, zeitgemäß umgesetzt, absolut funktionsfähig, technologisch perfekt gemacht, große Ausgießfläche in einem zweiten Material, was diesem Produkt eine erstaunliche Wertigkeit verleiht.

An archetypical Mediterranean form for drinks, modern, executed in a contemporary fashion, works outstandingly, technologically perfect, large pouring surface in a second material, making this an astonishingly high-quality product.

0 g
100 g
200 g
300 g
500 g
600 g
700 g
1 kg
eva solo

Hersteller ***Manufacturer***	**Vertrieb** ***Distributor***	**Design** ***Designer***
Eva Denmark A/S	Eva Deutschland	Claus Jensen & Henrik Hoelbæk
Hoejnaesvej 59	Pinneberger Chaussee 99	Tools Design
DK-2610 Roedovre	D-25436 Moorrege	Rentemestervej 23 A
Tel. +45 36 73 20 60	Tel. +49 41 22 8 18 62	DK-2400 Kopenhagen NV
www.evadenmark.com	Fax +49 41 22 8 92 22	Tel. +45 38 19 41 14

Küchenwaage „Eva Solo" *kitchen scales*

Eine traditionelle Küchenwaage besteht normalerweise aus etwa 35 Kunststoff- und Metallteilen oder noch mehr, wenn elektronische Komponenten dabei sind. Nicht so bei der Küchenwaage „Eva Solo". Sie besitzt gerade mal drei Teile und ist damit wohl die schlichteste und einfachste Küchenwaage: ein Metallbecher mit Meßskala auf der Außenseite, ein Glasbehälter mit Gewichtsskala und eine Edelstahl-Feder. Diese drei Teile fügen sich zu einer perfekt funktionierenden Waage für Wiegegut bis zu einem Kilogramm zusammen. Die Waage hat auch eine Tarierfunktion. So kann man jedes Mal, wenn man neue Zutaten zugibt, auf Null stellen und weiter wiegen, ohne die Schale entleeren zu müssen. Außer wiegen kann man auch messen; denn der Metallbecher hat auf der Innenseite eine Deziliter-Einteilung. Somit kann dieser als normaler Meßbecher benutzt werden. Das einfache Gerät läßt sich, was wohl mit keiner anderen Küchenwaage möglich ist, in der Spülmaschine waschen.

Traditional kitchen scales are usually made up of roughly 35 plastic and metal parts, or indeed even more if electronic components are included. This is not the case with the "Eva Solo" kitchen scales. These scales comprise just three parts, making them probably the simplest and most straightforward of all kitchen scales: a metal jug with a measuring scale on the outside, a glass jug with a weighing scale, and a stainless steel spring. Taken together, these three parts harmonize to make perfectly functional scales with which to weigh items up to one kilogram. The scales can also be tared, and so every time new ingredients are added, they can be reset at zero and the next ingredients weighed without having to empty the bowl first. Apart from weighing, liquids can be measured too, as there is a decilitre scale on the inside of the metal jug. This means that it can be used as a normal measuring jug. Probably unlike any other kitchen scales, this simple appliance can be washed in a dishwasher.

Jury ***Judges panel***

Funktionalität bestehend aus drei Teilen. Selbsterklärendes Funktionsprinzip, Lowtech-Waage, vielseitige Anwendungen: Vase, Behälter, Meßbecher und last but not least sogar noch eine Waage.

Functionality in three simple parts. Self-explanatory function, low-tech scale, a variety of uses: vase, vessel, measuring jug and even, last not least, kitchen scales.

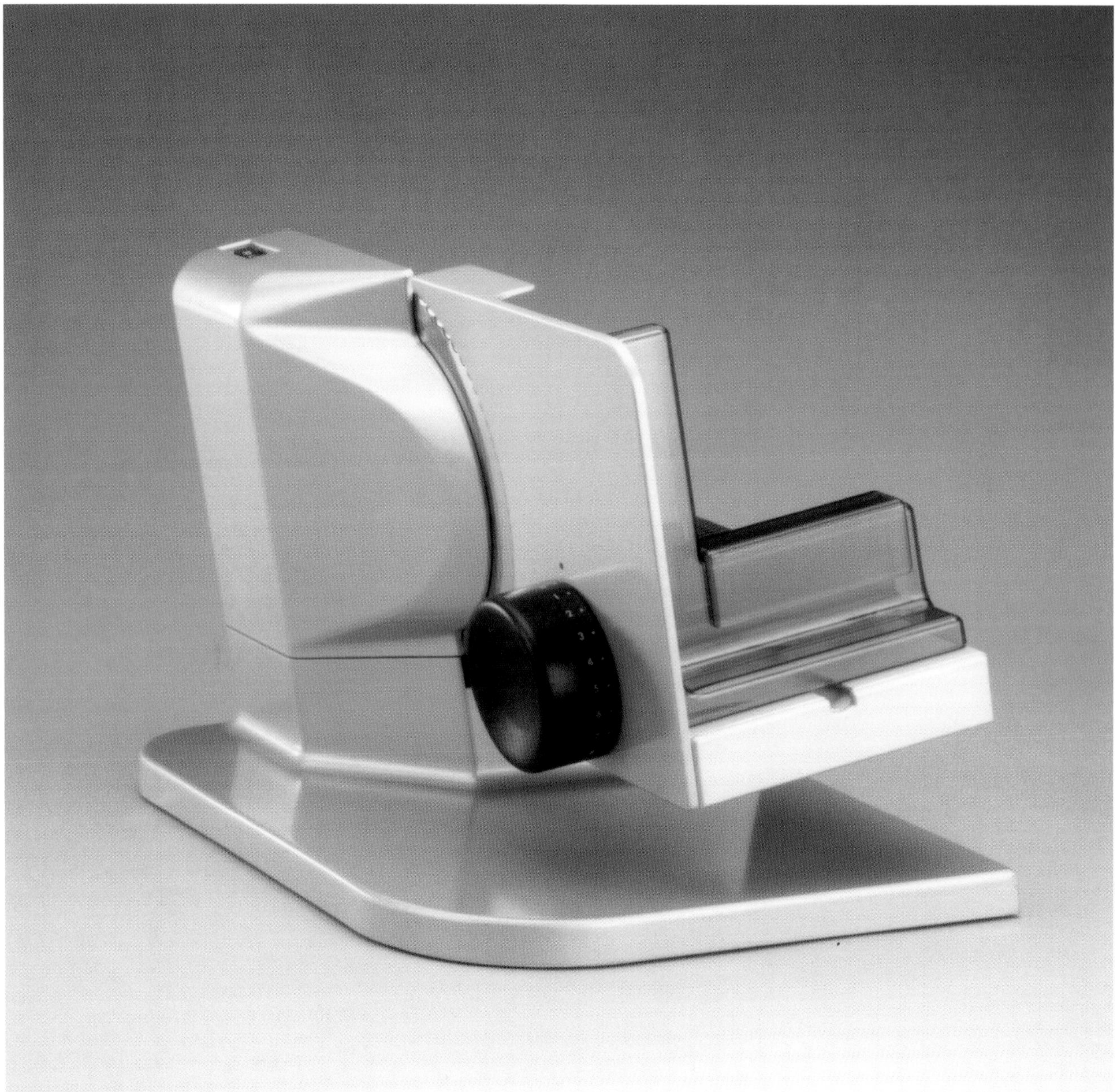

Hersteller *Manufacturer*	**Vertrieb** *Distributor*	**Design** *Designer*
Ritterwerk GmbH, Hausgeräte Industriestraße 13 D-82194 Gröbenzell Tel. +49 81 42 5 10 41 www.ritterwerk.de	Ritterwerk GmbH, Hausgeräte Industriestraße 13 D-82194 Gröbenzell Tel. +49 81 42 5 10 41 www.ritterwerk.de	Karl Dittert Eichenweg 32 D-73525 Schwäbisch Gmünd Tel. +49 71 71 6 22 63 Fax +49 71 71 6 12 60

Multischneider „E 21 Freischneider Diagonal" *all-purpose food slicer*

Unzerlegte Lebensmittel bieten viele Vorteile: sie bleiben länger frisch, sind oft preiswerter, können in größeren Mengen gekauft und aufbewahrt werden. Ein guter Allesschneider, wie der Multischneider, schneidet dann zu Hause vom Brot über Wurst und Fleisch bis zu Käse und Zwiebeln alles in dicke oder hauchdünne gleichmäßige Scheiben. Dank der bedarfsgerechten Konstruktion hat er einen hohen Gebrauchswert. So ist die freitragende Metallkonstruktion mit integrierter Bodenplatte besonders stabil und handlich und erlaubt das direkte Ablegen der geschnittenen Lebensmittel auf Teller und Platten. Die Schrägstellung des Gehäuses – von Auflegetisch, Andruckplatte und Messer – garantiert beste Schneideergebnisse besonders bei sehr dünnem Schneidgut, das nicht „in sich" zusammenfällt, sondern sich zu appetitlichen Scheiben „rollt". Das Gerät zeichnet sich durch anspruchsvolles, funktionales Design aus. Dank der übersichtlichen Anordnung der Bedienelemente läßt es sich sicher und einfach handhaben. Der Moment- und Dauerschalter ist kindersicher. Die gute Material- und Verarbeitungsqualität macht das Produkt langlebig: Metall und hochwertiger Kunststoff, alles silbermetallic.

There are many advantages of buying food uncut: it stays fresh longer, it is often less expensive, and it can be bought and kept in larger quantities. With a good all-purpose food slicer like the "Multischneider", anything from bread via sausage and meat to cheese and onions can be cut into thick or wafer-thin slices at home. Designed with users' needs in mind, it is a pleasure to use. The cantilevered metal design with its integrated base plate is particularly robust and easy to manage, and allows the food that has been sliced to be laid directly onto plate or serving dish. The angled housing – and with it the feedboard, guard plate and slicer – guarantees optimum slicing results, especially if the food is being cut very thin. It will not crumble and collapse, but "roll" into appetizing slices. The appliance has a sophisticated, functional design. The control elements have been arranged clearly, making it safe and easy to use. The cut-out switch and switch for permanent operation is childproof. Good quality materials and finishing make this a long-lasting product: metal and high-quality plastic, everything in a silver-metallic finish.

Jury ***Judges panel***

Kompakte Hausmannskost, das heißt ein Gerät mit schöner Mechanik auf kleinstem Raum effektiv gemacht, benutzerfreundlich, schöne Proportionen, optimale Abmessungen. Hohe Wertigkeit in der Materialgestaltung und in der Materialkonfrontation. Eine andere Ästhetik, es ist nicht mehr die Fleischermaschine, die man aus den Wurstläden kennt.

Compact plain fare, i.e. an appliance with a good mechanism taking up a minimum of space, effectively made, user-friendly, nicely proportioned, optimum dimensions. Very high-quality in the design of the materials and the way they have been confronted with each other. Aesthetically different: no longer the cutting machine you find in butcher's shops or delicatessen counters.

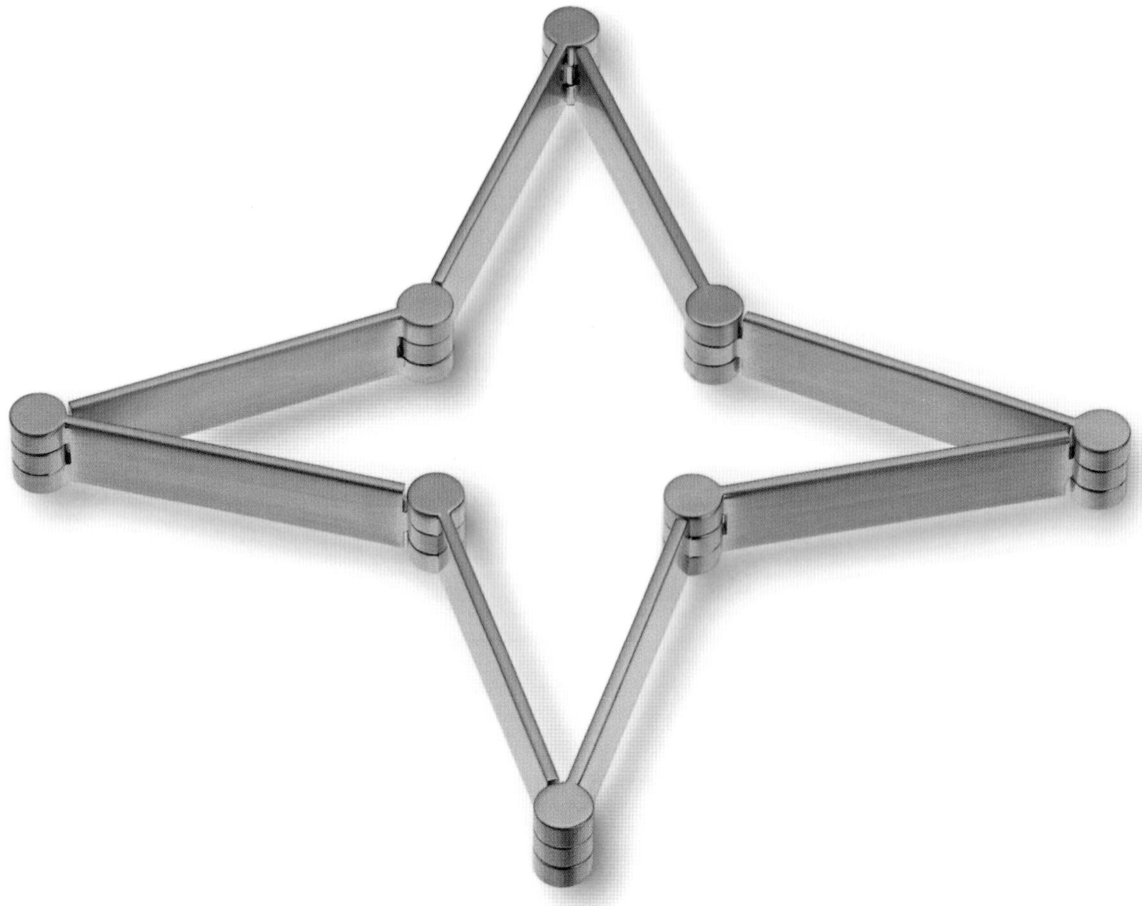

Hersteller ***Manufacturer***	**Vertrieb** ***Distributor***	**Design** ***Designer***
Auerhahn Bestecke GmbH Postfach 1161 D-72206 Altensteig Tel. +49 74 53 94 68 34 www.auerhahn-bestecke.de	Auerhahn Bestecke GmbH Postfach 1161 D-72206 Altensteig Tel. +49 74 53 94 68 34 www.auerhahn-bestecke.de	Vogt & Weizenegger Mulackstraße 15 D-10119 Berlin Tel. +49 30 6 14 21 66 Fax +49 30 6 14 22 87

Topfuntersetzer „A-Design" *trivet*

Mal ist der Topfuntersetzer ein Quadrat, dann ein Stern oder ein Achteck. Dank seiner acht Gelenke paßt er sich jeder beliebigen Form an, so daß man ihn als Untersetzer für die verschiedensten Töpfe und Schüsseln einsetzen kann. Er eignet sich beispielsweise in länglichen Formen für Servierplatten oder Bräter, in kompakten Formen für Töpfe oder Schüsseln. Der Topfuntersetzer ist aus mattem Edelstahl hergestellt.

A trivet that is now a square, now a star, or now an octagon. Its eight joints allow it to take on any form, and so it can be used as a stand for a wide variety of saucepans or bowls. In a more oblong form, for example, it is suitable for serving platters or roasting tins, while its compact form is better for pans or bowls. The trivet is made of matt stainless steel.

Jury ***Judges panel***

Und schon wieder ein Untersetzer, dieses Mal mit hoher Präzision und klassischer Tradition, durch intelligente spielerische Veränderung unterschiedliche Flächen zu bedienen.

Yet another trivet, but this time one made with great precision. Part of a classical tradition of catering for different surfaces by intelligent, playful changes of form.

Hersteller ***Manufacturer***	**Design** ***Designer***
Brabantia Brabantialaan 3 NL-5583 BP Waalre Tel. +31 40 2 28 22 58 www.brabantia.com	Brabantia Brabantialaan 3 NL-5583 BP Waalre Tel. +31 40 2 28 22 58 www.brabantia.com

Abfallsammler „Twin bin" *waste bin*

„Twin bin" gehört zu dem innovativen Touch bin® Konzept der Firma. Eine leichte Berührung, der Deckel öffnet sich und der Müll verschwindet. Restmüll und kompostierbare Abfälle können getrennt gesammelt werden. Dafür gibt es zwei Inneneimer aus Polypropylen: einen 23-Liter-Eimer für den Restmüll und einen 10-Liter-Eimer für den kompostierbaren Müll. Sinnvolle Details steigern den Komfort. So kann dank spezieller Lüftungslöcher im großen Inneneimer überflüssige Luft, die sich beim Einlegen des Müllbeutels aufgebaut hat, entweichen. Die extra große Öffnung des Abfallbehälters verhindert, daß etwas darüber hinausfällt. Außerdem läßt er sich mühelos reinigen. Der solide Tragegriff kann leicht bewegt werden, selbst wenn der Mülleimer voll ist. Der Boden des Abfallsammlers ist aus Polypropylen und schützt somit den Fußboden vor Kratzern und Beschädigungen. Der entfernbare Deckel besteht aus Edelstahl, das robuste Gehäuse aus hochwertigem Chromstahl. Die korrosionsbeständigen Materialien und das ovale Design heben den Abfallsammler von anderen ab. „Twin bin" zeichnet sich durch eine solide, funktionelle und elegante Gestaltung aus. Das langlebige Produkt hat eine zehnjährige Garantie.

"Twin bin" belongs to the company's innovative Touch bin® line. A light touch is all that is needed for the lid to open and the waste to disappear. Residual and compostable waste can be collected separately, thanks to two interior bins made of polypropylene: a 23-litre bin for residual waste and a 10-litre bin for compostable waste. Sensible details make it more convenient to use. Special ventilation holes in the large internal bin, for example, allow air to escape when a bin liner is inserted . The waste bin's extra large opening prevents anything falling over the side. Moreover, it can be cleaned effortlessly. The robust carrying handle can easily be moved even if the waste bin is full. Because the bottom of the waste bin is made of polypropylene, it does not scratch or damage the floor. The removable lid is made of stainless steel, and the robust housing of high-quality chrome steel. The rustproof materials and oval design set this waste bin apart from others. "Twin bin" has a robust, functional and elegant design. This long-lasting product has a ten-year guarantee.

Jury ***Judges panel***

Eleganter zeitloser Müllbehälter mit guter Sammelfunktion für Feuchtmüll und normalen Müll, funktional auch dahingehend, daß der Feuchtmüllbehälter dementsprechend kleiner gemacht wurde, um eine frühere Leerung zu erzwingen. Bestechend solide, überraschend gut gemacht.

Elegant, timeless waste bin, works well for the collection of wet and normal waste. Also functional in that the container for wet waste is smaller, making it necessary to empty it earlier. Strikingly robust, surprisingly well made.

Hersteller *Manufacturer*	**Vertrieb** *Distributor*	**Design** *Designer*
Casco Schutzhelme GmbH Gewerbering Süd 11 D-01900 Bretnig-Hauswalde Tel. +49 3 59 55 83 90 www.casco-schutzhelme.de	Waldhausen GmbH & Co. KG Von-Hünefeld-Straße 53 D-50829 Köln-Ossendorf Tel. +49 2 21 58 80 10 Fax +49 2 21 5 88 01 25	Werksdesign

Reithelm „Duell" *riding cap*

„Duell" ist ein Profi-Helm nicht nur für den Turniersport sondern auch für engagierte Freizeitreiter. Er erfüllt hohe Ansprüche an Komfort, Sicherheit und Design. Seine Eleganz, seine äußerst schlanke Formgebung und sein unsichtbares, aber hocheffektives Belüftungssystem charakterisieren dieses Modell. Der perlschwarze Helm ist mit einem doppelten Monocoque Seitenschutz ausgestattet. Die Mischung verschiedener Materialien gibt ihm nicht nur sein besonderes Aussehen, sondern bietet eine zusätzliche passive Sicherheit: Stoffeinsätze und Schild reflektieren bei einsetzender Dämmerung im Scheinwerferlicht. Dank der neuen Kordelberiemung kann man den Kinngurt – ohne den Helm abzunehmen – jederzeit straffen und lockern.

"Duell" is a cap for the expert, not only for tournaments but also for enthusiasts who ride a lot in their spare time. It meets high standards of comfort, safety and design. The model is characterized by elegance, an extremely slim shape and a concealed but highly effective ventilation system. The pearl-black cap is equipped with double monocoque side protection. The mixture of different materials used not only gives it its special appearance, but also provides additional passive security: The fabric inserts and visor act as reflectors during twilight, or in the glare of headlights. The new cord strap allows the chin strap to be tightened and loosened at any time without taking off the cap.

Jury ***Judges panel***

Ein Hightech-Produkt, das in der Materialwahl die klassische Reiter-Kopfbedeckung zitiert. Eine sehr leichte Version, die dem Reithelm das Monströse genommen hat. Atemfreundlich. Sehr schön die Zwei-Materialien-Komponente im Kunststoffbereich: glänzend und matt. Eine Allegorie an das klassische Material Samt, das heute synthetisch ist. Sehr feine Innenverkleidung.

A high-tech product whose choice of materials quotes the classical riding cap. A very light version that has done away with the cumbersome aspects of a riding cap. Well ventilated. One very attractive feature is the two types of synthetic materials used: glossy and matt. An allegory of the classical velvet material, which today is synthetic. Very refined inner lining.

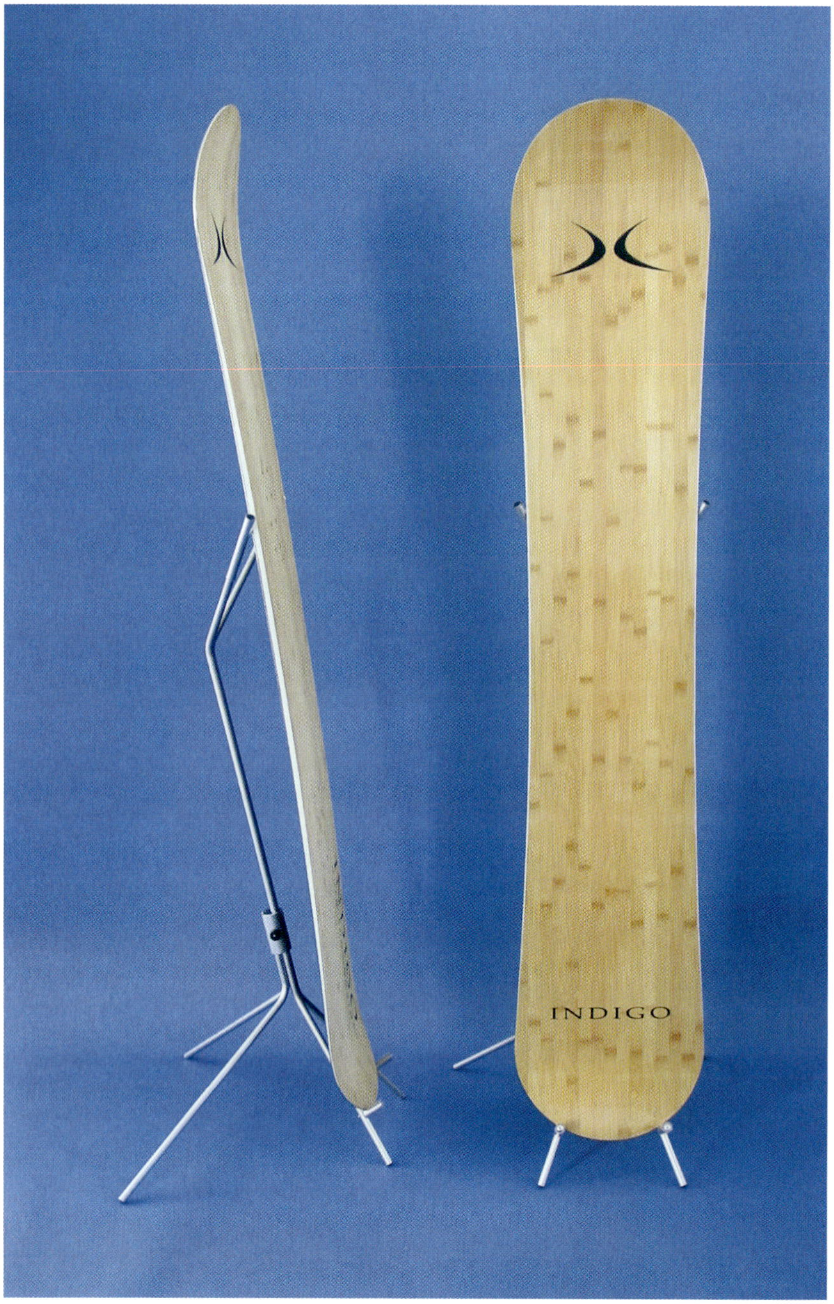
INDIGO
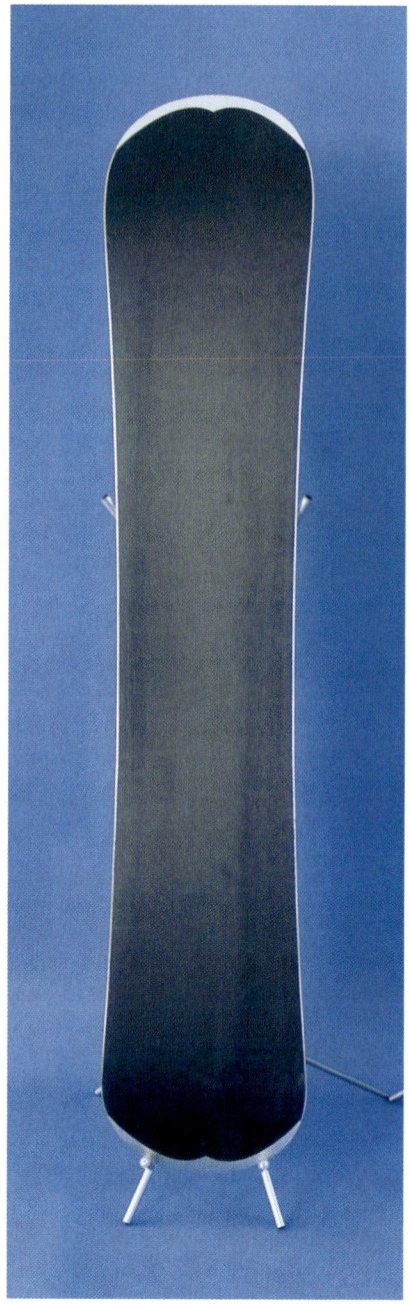

Hersteller ***Manufacturer***	**Vertrieb** ***Distributor***	**Design** ***Designer***
Indigo Pure Snowboards Thorsten Schwabe & Gregor Baer Blutenburgstraße 41a D-80636 München Tel. +49 89 18 95 45 50 www.indigosnow.de	Indigo Pure Snowboards Thorsten Schwabe & Gregor Baer Blutenburgstraße 41a D-80636 München Tel. +49 89 18 95 45 50 www.indigosnow.de	Gregor Baer, Thorsten Schwabe Werksdesign

Snowboard „Indigo Freeride 163“ *snowboard*

Das Snowboard „Indigo Freeride 163“ ist eine Weltneuheit, da es aus Bambus gefertigt wurde. Es entspricht in Material, Form und Konstruktion den hohen Anforderungen, die an ein Freeride-Snowboard gestellt werden. Die Designer sind selbst mit dem Snowboardsport aufgewachsen, was sich auf ihre Gestaltung der Snowboards auswirkt. Die Firma hat es sich zur Aufgabe gemacht, hochwertige Freeride-Snowboards aus reinem Holz zu entwickeln, die in jedem Gelände unter härtesten Bedingungen funktionieren. Das neu entwickelte Bambusmodell ist handgefertigt. Es besteht aus Echtholz-Bambusfurnier, das stoßfrei verpreßt wurde. Vorgänger waren Snowboards mit Oberflächen aus Eschen- und Wurzelholz. Der eigens entwickelte Vollholzkern wird durch fein abgestimmte Karbongewebe in Ober- und Untergurt zu einem aggressiven Hochleistungssportgerät. Das Snowboard „Indigo Freeride 163“ besticht nicht nur durch sein zeitlos schlichtes Erscheinungsbild sondern auch durch seinen hohen Anspruch an Funktionalität und Stabilität in einer Bauweise, die das Gewicht optimiert. Durch die hauseigene Produktion ist es möglich, auch individuelle Designs in Kleinststückzahlen zu fertigen.

The "Indigo Freeride 163" is the first snowboard of its kind in the world, because it is made of bamboo. Materials, shape and design satisfy the high demands made of a freeride snowboard. The designers themselves grew up with snowboard sport, which is reflected in the design of their snowboards. The company has set itself the task of developing high-quality freeride snowboards out of pure wood, which work on any kind of slope under extreme conditions. The newly developed bamboo model is hand made. It is made up of real bamboo veneer that has been compressed without joints. Predecessor models had surfaces made of ash and root timber. Combining the specially developed all-wood core with specifically selected woven carbon textile on the top and bottom surfaces results in an aggressive piece of high performance sports equipment. The "Indigo Freeride 163" snowboard not only stands out as a result of its timelessly unadorned appearance, but also as a result of the exacting demands it places on function and stability in a constructional form that optimizes weight. Because they are produced in the company, small batches can also be produced in individual designs.

Jury ***Judges panel***

Das Produkt überrascht durch seinen Materialeinsatz: das traditionelle Material Bambus für ein Fun-Sport-Gerät. Eine amüsante Variante im jugendlichen Chic über das Bambusmaterial aufgewertet, spannend.

The use of materials in this product is astonishing: bamboo, a traditional material, used for an item of fun-sport equipment. An amusing variant in youthful chic, given a higher quality by the bamboo, exciting.

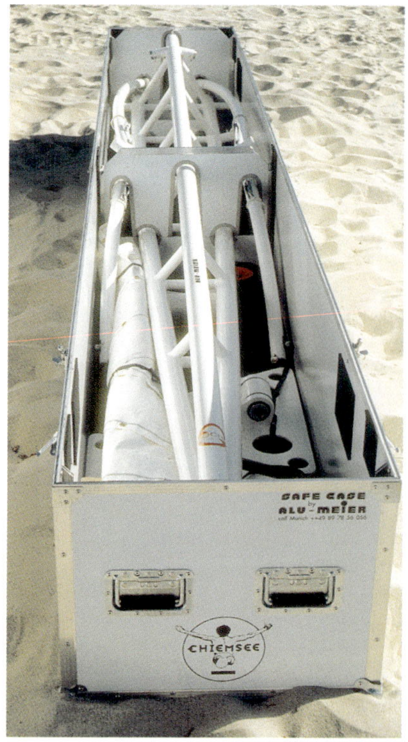
SAFE CASE
ALU-MEIER
CHIEMSEE

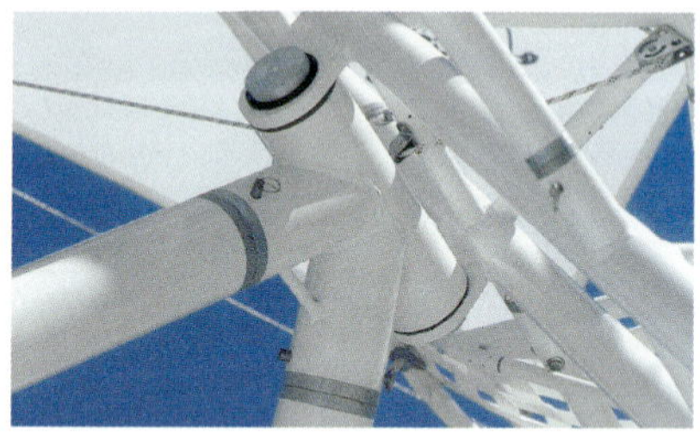

Hersteller ***Manufacturer***	**Design** ***Designer***
Alu Meier Kistlerhofstraße 140 D-81379 München Tel. +49 89 7 48 90 50 www. alu-meier.de	Jürgen Schubert, Thorsten Schwabe, Peter Zimmer Mitarbeit: Marcus Kottermair Blutenburgstraße 41a D-80636 München Tel. +49 89 18 95 45 41

Beobachtungsturm „Beach Point®" *observation tower*

„Beach Point" ist ein multifunktionaler Beobachtungsturm. So kann er nicht nur als mobile Wasserwachstation sondern auch als Aussichts-, Schieds- und Signalturm für Wellenreitwettbewerbe, Segel- und Surfregatten eingesetzt werden. Dafür ist er mit zwei Hochsitzen mit verstellbarer Verschattung ausgestattet. Außerdem bietet „Beach Point" eine Liegefläche mit Sonnensegel. Dank seines außergewöhnlichen Designs wird er zum werbewirksamen Eyecatcher und unverwechselbaren Erkennungszeichen des Strandes, so daß er auch als Treffpunkt für Gemeinschaftsaktivitäten dient. Seine leichte filigrane Gestaltung sticht ins Auge. „Beach Point", bei dem auch bewährte Materialien und Detaillösungen aus dem Segelbootbau angewendet wurden, ist eine pulverbeschichtete Aluminium-Leichtbaukonstruktion mit einer Gesamthöhe von 11 Metern. Textile Membranen bilden Dach, Liegeflächen, Sonnensegel und Zelt. Die leichten und handlichen Einzelbauteile aus hochbeanspruchbaren Materialien lassen sich problemlos und leicht von zwei Personen nur mit einem Hammer innerhalb einer Stunde montieren. Er ist leicht verstaubar und auch an einem schwer zugänglichen Standort aufstellbar.

"Beach Point" is a multifunctional observation tower. Not only can it be used as a lifeguard's mobile watchtower, but also as a viewing, refereeing or signal tower for surfing competitions, sailing and windsurfing regattas. For this purpose, it is equipped with two high chairs, with variable sunshades. "Beach Point" also has a reclining surface with a sunshade. Its extraordinary design means that it is an eye-catching feature that can be used to great effect for advertising purposes, as well being an unmistakable landmark on the beach, making it ideal as a meeting point for communal activities. Its light, filigree design is very striking. "Beach Point", in which materials and ideas from sailboat construction have also been used, is a powder-coated lightweight aluminium construction with an overall height of 11 metres. Textile membranes form the roof, reclining surfaces, sunshades and tent. The light, easy to handle individual components, made of hard-wearing materials, can be put together with no difficulty by two persons in less than an hour. The only tool they need is a hammer. It can easily be stowed away and can also be erected in places where access is difficult.

Jury ***Judges panel***

Ein altes Objekt wurde komplett neu überarbeitet, symbolisiert den Einsatzort mit Luftigkeit, Wind, Leichtigkeit. Ermöglicht, daß mehrere Leute sich auf der katamaranähnlichen Trapez-Liegewiese sonnen können. Abbaubar, transportier- und verstaubar für Wintermonate, dann wieder nur mit geringem Zeitaufwand aufbaubar.

A familiar object has been completely rethought. The places it is used are symbolized in its airiness, wind, lightness. Several people can lie in the sun on its catamaran-like trapeze surfaces. Can be knocked down, transported and stowed away in the winter months, then set up in a minimum of time.

S
class

S class
uwe

Hersteller ***Manufacturer***	**Vertrieb** ***Distributor***	**Design** ***Designer***
Uwe GmbH Buchstraße 82 D-73525 Schwäbisch Gmünd Tel. +49 71 71 10 32 16 info@uwe-germany.de	Uwe GmbH Buchstraße 82 D-73525 Schwäbisch Gmünd Tel. +49 71 71 10 32 16 info@uwe-germany.de	Ottenwälder und Ottenwälder Sebaldplatz 6 D-73525 Schwäbisch Gmünd Tel. +49 71 71 92 72 30 www.ottenwaelder.de

Solarium „S-class" *solarium*

Man wollte ein Solarium für höchste Ansprüche entwickeln, das nicht nur technologisch neue Maßstäbe setzt, sondern auch eine Oase für Körper, Geist und Seele bietet. So wurde das Design darauf ausgerichtet, den Körper zu entspannen und die Sinne zu berauschen mit einem perfekt aufeinander abgestimmten neuentwickelten Konzept aus Sonne, Wärme, Wasser und Duft. Innovativ und einzigartig ist dabei das neue „Aroma". Hier werden während dem Sonnenbaden in unterschiedlichen Phasen Düfte freigesetzt. Jede Phase wirkt anders: relaxend, beruhigend, stimulierend, vitalisierend. Mit „Breeze" wird jetzt aus zwei Düsen die Haut besonders intensiv und gleichmäßig von kühlem Wassernebel benetzt und erfrischt. Bei dem Controlled Power System CPS sorgt das naturnahe Verhältnis von UVA und UVB für eine intensive, langanhaltende Pigmentierung. Durch Zu- und Abschalten einzelner Lampen kann die Intensität individuell gesteuert werden. Das innovative Farbkonzept basiert auf speziellen im Licht changierenden Metalliclacken, wie sie von Automobilveredlern verwendet werden. Das Design zeichnet sich durch eine weiche, beschützende Formensprache mit kraftvollen dynamischen Akzenten aus.

The idea was to develop a solarium for the most discerning user. Not only a solarium that sets new technological standards, but also an oasis for body, mind and soul. The design has therefore been geared to relaxing the body and exhilarating the senses with a perfectly matching and coordinated new concept made up of sun, warmth, water and fragrance. One unique, innovative feature is the new "aroma". During the different phases of sun-bathing, fragrances are released. Each phase has a different effect: relaxing, calming, stimulating, vitalizing. The "breeze" feature wets and refreshes the skin intensively and evenly with a cool mist emitted by two nozzles. The CPS Controlled Power System ensures that the ratio of UVA to UVB is practically the same as in normal sunlight, ensuring an intensive, long-lasting tan. Lamps can be switched on and off in isolation, allowing intensity to be controlled individually. The innovative colour scheme is based on special metallic paint that changes its colour in the light, similar to that used by car paint shops. The formal expression of the design is one of softness and protection, to which powerful, dynamic accents have been added.

Jury ***Judges panel***

Keine einfache Entscheidung für die Jury, aber unter dem Dach Lebensart ist das Wellness- und Schönheitsthema nicht zu ignorieren. Diese Sonnenbank nimmt das Wohlbefinden des sich Bräunenden außerordentlich ernst und versucht neben medizinischen Aspekten auch die Wohlfühlaspekte zu berücksichtigen. Gleichzeitig ist es eine Herausforderung, das Design weiterzuentwickeln.

No easy decision for the judges, but the "way of life" remit also means that the wellness and beauty area cannot be ignored. This sun bed takes the well-being of the sunbather very seriously, attempting to consider both medical features and wellness aspects in the way it is used. At the same time, the design is a challenge to pick up from where this sunbed leaves off.

HOHNER
ventura IV 120

concerto IV 120

HOHNER

Hersteller ***Manufacturer***	**Vertrieb** ***Distributor***	**Design** ***Designer***
Matthias Hohner AG Hohnerstraße 8 D-78647 Trossingen Tel. +49 74 25 2 00 Fax +49 74 25 2 03 71	Musik Meyer GmbH Industriestraße 20 D-35041 Marburg Tel. +49 64 21 98 90 Fax +49 64 21 98 93 59	Bluequest Design Group und SquareOne GmbH (früher Topel und Pauser Industrial Design)

Akkordeon „Ventura IV 120" *accordion*

„Ventura IV 120" ist ein Akkordeon in völlig neuer zeitgenössischer Optik. Bei ihm verbindet sich modernes hochwertiges und zugleich zeitloses Design mit innovativer Technik. So wurde bei „Ventura" eine völlig neu entwickelte Tastatur eingesetzt, mit der bereits bei Highclass-Instrumenten sehr gute Erfahrungen gemacht wurden. Dank ihren nahezu reibungsfreien Führungselementen hat sie einen sehr leichten Tastenlauf. Denselben hohen Anspruch erfüllt auch die neu gezeichnete Baßmechanik. Für weitere komfortable Handhabung für den Akkordeonist sorgen die ergonomischen Detaillösungen wie die sehr vielfältigen Riemenaufnahmen und die stufenlos einstellbaren Trageriemen „Ergoline". Das Akkordeon hebt sich von anderen durch seine moderne, elegante Gestaltung ab. Zu dem Gehäuse in edlem Schwarz bilden die silbermetallic schimmernden Verdeckteile einen schönen Kontrast. Ins Auge sticht auch die konkav geschwungene Linie, die das Verdeck über das Diskantgehäuse zieht und einen harmonischen Abschluß am Balg findet.

"Ventura IV 120" is an accordion with a completely new, contemporary appearance. It combines modern, high-quality yet timeless design with innovative technology. A completely new keyboard development has been used in "Ventura", for example, which has already proved its worth in high-class instruments. Thanks to its almost friction-less guide elements, the keys can be run over very easily. The newly designed bass keyboard satisfies the same high criteria. Other ergonomically designed details make the instrument easy to handle, such as the many different places to which straps can be attached and the infinitely adjustable "Ergoline" carrying strap. This accordion stands out from others by virtue of its modern, elegant form. The gleaming silver-metallic hood parts contrast beautifully with the elegant black housing. A further striking element is the concave line that the hood takes over the treble keyboard housing, ending harmoniously at the bellows.

Jury ***Judges panel***

Nostalgie-Gerät mit hoher Aktualität. Schöne Umsetzung traditioneller Materialien in zeitgenössischer Technologie, also Kunststoff statt Aluminium. Sehr subtil gelöst. Beeindruckend der offensichtliche Qualitätsaspekt bei diesem Instrument.

Nostalgic instrument that is as up to date as ever. Nice combination of traditional materials with contemporary technology – plastic instead of aluminium. Very subtle solution. The obvious quality aspect in this instrument is impressive.

Hersteller ***Manufacturer***	**Vertrieb** ***Distributor***	**Design** ***Designer***
Baumann + Trapp GbR Mattenmühle 1 D-77876 Kappelrodeck Tel. +49 78 41 29 03 00 www.ideen-fuer-kinder.com	Baumann + Trapp GbR Mattenmühle 1 D-77876 Kappelrodeck Tel. +49 78 41 29 03 00 www.ideen-fuer-kinder.com	Dieter Baumann, Wilfried Trapp Werksdesign

Balancekombinationsspiel „Bärlino" *balancing toy*

„Bärlino" ist eine Weiterentwicklung des großen, einen halben Meter langen Handbalancierspiels „Murmelbär", das in Kindergärten inzwischen zum Klassiker geworden ist. Man wollte ein Reisehandspiel entwerfen, das gleichzeitig viele Funktionen integriert und verschiedene Altersgruppen anspricht. Entstanden ist ein Kombinationsspiel, das die Sinne anregt, die Feinmotorik und den Gleichgewichtssinn spielerisch fördert und viel Spaß macht – ein idealer Reisebegleiter für Kinder. Es gibt rasend schnelle Achterbahnen mit Hohlnut, damit die Murmeln nicht herausspringen können. Außerdem bietet „Bärlino" Lupe, Kompaß und Thermometer. So können auf der Reise Landkarten, Mücken, Steine oder Muscheln betrachtet, Himmelsrichtungen geraten und die Temperatur geschätzt und gleich kontrolliert werden. Daß alle Module in den Bären aus MDF integriert sind, bietet einen weiteren Vorteil: die Kleinteile werden im täglichen Gebrauch nicht leicht verlegt oder verloren, womit der übliche „Such-Ärger" bei Kindern vermieden wird.

"Bärlino" is a further development of the large, 50 centimetre long "Murmelbär" hand-held balancing toy, which has meanwhile become a kindergarten classic. The idea was to develop a travel toy which integrates many functions simultaneously and appeals to various age groups. What has emerged is a combination toy that stimulates the senses, playfully supports the development of fine motor skills and the sense of balance, and is a lot of fun – an ideal travel companion for children. There are high speed hollow-section figures of eight, which the marbles cannot fall out of. "Bärlino" also features a magnifying glass, compass and thermometer. As they journey along, children can study maps, flies, stones or sea shells, guess directions and the temperature, and check their answers straight away. As all the features are integrated in the MDF bears, there is one further advantage: the small parts cannot be mislaid or lost in daily use, and so the annoyance of having to look for children's things is avoided here.

Jury ***Judges panel***

Ein Spielzeug von hoher grafischer und haptischer Qualität. Pädagogischer Ansatz mit Lernmöglichkeiten. Bestechend die technische Ausführung, zum Beispiel, daß die Kugeln nicht herausfallen durch die hinterschnittene Nut. In der Produktionsweise Hightech, aber im Erscheinungsbild Lowtech.

A toy with excellent graphic and tactile qualities. Educational approach with the possibility to learn. The technical design is excellent: for example, the way the hollow section has been cut means that the marbles cannot fall out of the groove. A high-tech production method, but a low-tech appearance.

Hersteller ***Manufacturer***	**Vertrieb** ***Distributor***	**Design** ***Designer***
Baumann + Trapp GbR Mattenmühle 1 D-77876 Kappelrodeck Tel. +49 78 41 29 03 00 www.ideen-fuer-kinder.com	Baumann + Trapp GbR Mattenmühle 1 D-77876 Kappelrodeck Tel. +49 78 41 29 03 00 www.ideen-fuer-kinder.com	Dieter Baumann, Wilfried Trapp Werksdesign

Balancehandspiel „Minischeiben" *balancing toy*

Hinter dem Entwurf des Minibalancehandspieles steht folgende Beobachtung: Die meisten kleinen Balancierspiele mit Kugeln werden bei Kindern schnell uninteressant, weil die Kugeln häufig auf den Boden fallen und dort schwierig zu finden sind. Sind außerdem noch kleinere Geschwister da, besteht die Gefahr, daß sie die kleinen Kugeln verschlucken. Deshalb wurden bei diesem Balancespiel die Nuten der Schleifen als Hohlnut ausgebildet, so daß die Stahlkugeln nicht herausfallen können. Das Spiel schult spielerisch Koordination, Konzentration und Sinneswahrnehmung. Außerdem fördert es gleichzeitig das Balancegefühl, trainiert die Feinmotorik der Hände und lockert die Gelenke. Dank der anderen Nutform kann neben den bekannten Geschicklichkeitsspielen zusätzlich auch ein Balancierspiel über dem Kopf ausprobiert werden, was ganz andere Anforderungen an den Gleichgewichtssinn stellt. Von dem Minihandspiel aus MDF gibt es zwei verschiedene Ausführungen. Beim roten Labyrinth gewinnt, wer die Kugel am schnellsten in die Mitte bringt. Bei Zickezacke in gelb läßt man die Kugel vom Ausgangs- bis zum Endpunkt rollen.

The observation that prompted the design of this miniature balancing toy was that children quickly lose interest in most small balancing toys with marbles or balls in them because the marbles often drop out and are difficult to find on the floor. Moreover, if small brothers or sisters are around, there is the danger that they will swallow the small marbles. That is why the loops in this balancing toy have been designed as hollow sections, so that the steel balls cannot fall out. This toy playfully trains coordination, concentration and sensual perception. It also supports the development of a sense of balance, trains manual fine motor skills and creates supple wrists. Apart from the usual dexterity games, the hollow-section groove means that the child can try balancing over its head, which makes quite different demands of its sense of balance. There are two models of the miniature MDF toy. In the red maze version, the winner is whoever gets the ball to the centre quickest. In the yellow zig-zag, the idea is to roll the ball from the start to the finish.

Jury ***Judges panel***

Bestechende Materialtechnologie. Endlich mal wieder etwas, das nicht aus Kunststoff ist. Übt mit Sicherheit auf Kinder über Grafik und bewegende Teile eine Faszination aus. Die Entwicklung motorischer Fähigkeiten wird gefördert, als Alternative zum Gameboy. Digitale versus manuelle, handwerkliche Geräte. Kann auf dem Rücken liegend gespielt werden.

Fascinating materials technology. At last something that is not made of plastic. Certain to fascinate children with its graphics and moving parts. Encourages the development of motor skills, as an alternative to the GameBoy. Digital versus manual, hand-made devices. Can be played while lying on your back.

Hersteller ***Manufacturer***	**Vertrieb** ***Distributor***	**Design** ***Designer***
Naef Spiele AG Grendelweg 5 CH-4314 Zeiningen Tel. +41 61 8 51 18 44 www.naefspiele.ch	Naef Spiele AG Grendelweg 5 CH-4314 Zeiningen Tel. +41 61 8 51 18 44 www.naefspiele.ch	Heiko Hillig Dorfstraße 68 D-09518 Mauersberg Tel. +49 1 71 2 35 54 54 Fax +49 37 35 9 08 62

Mosaikbild „Imago" *mosaic picture*

„Imago" besteht aus 36 exakt verarbeiteten Ahornwürfeln, die in einen umlaufenden großzügigen Rahmen gesteckt werden. Jeder einzelne Würfel besitzt sechs grafische Varianten: verschiedene Muster aus schwarz-weißen Punktrastern sowie eine Naturoberfläche und eine schwarze Seite. Mit diesen Würfeln lassen sich unzählige abstrakte Bilder gestalten. Sowohl Kinder als auch Erwachsene können hier spielerisch mit Flächen umgehen und immer wieder völlig neue Varianten bauen, die an die Wand gehängt werden können und einen schönen Blickfang bilden. Hierbei sind der Phantasie und der Kreativität keine Grenzen gesetzt. Wer will, kann sich damit täglich an einem neuen Bild an der Wand erfreuen.

"Imago" consists of 36 precisely machined maple cubes, which are inserted in a wide-rimmed frame. Each individual cube offers six graphic variants: various black-and-white pointillist patterns, a natural surface and a black surface. Countless abstract pictures can be formed using these cubes. Here, both children and adults can experiment playfully with surfaces, composing completely new variations over and over again, which can be hung on the wall and form an attractive focal point. There are no limits to fantasy or creativity. If so desired, a new picture on the wall can delight the eye every day.

Jury ***Judges panel***

Bestechend die Kontinuität eines Unternehmens, das sich seit Jahrzehnten mit pädagogischem Spielzeug beschäftigt und das in der Regel auf höchstem qualitativem Niveau macht.

The striking thing here is the continuity of a company that has been in the educational toy business for decades, and generally produces on the highest qualitative level.

lemonfish

Hersteller ***Manufacturer***	**Vertrieb** ***Distributor***	**Design** ***Designer***
Lemonfish Alexandra Dittrich + Bettina Weitershagen GbR Lindenfirststraße 14/1 D-73525 Schwäbisch Gmünd www.lemonfish.de	Lemonfish Alexandra Dittrich + Bettina Weitershagen GbR Lindenfirststraße 14/1 D-73525 Schwäbisch Gmünd www.lemonfish.de	Alexandra Dittrich, Bettina Weitershagen Werksdesign

Tasche „4in1" *handbag*

„4in1" – die Handtasche mit vier Möglichkeiten. Je nach Stimmung, Tageszeit, Anlaß oder Kleidung wählt man aus vier verschiedenen Farbkombinationen aus und bekommt jedes Mal eine andere Tasche: schwarz-schwarz, schwarz-beige, schwarz-apfelgrün und schwarz-grau. Das ist möglich, weil Vorder- und Rückseite sowie die Klappenseiten unterschiedliche Farben haben. Will man die Farbe ändern, wird einfach die Taschenklappe umgeklappt oder die Tasche gedreht. Eine moderne zeitgemäße Tasche, vielseitig einsetzbar, mit vielen Variationsmöglichkeiten, die die Lust am Verändern und am Ausdruck des persönlichen Lebensstils befriedigt. Die Tasche zeichnet sich durch innovatives Design, viele praktische Details und hochwertige Materialien aus. Zwei große Hauptfächer, eine intelligente Inneneinteilung und viele nützliche Taschen für Kleinigkeiten bieten Stauraum für die Dinge des Tages.

"4in1" – the handbag with four faces. Depending on mood, time of day, occasion or clothing there is a choice of four colour combinations, each of them creating a different handbag: black-black, black-beige, black-apple green and black-grey. This is made possible by the different colours of the front and rear panels, as well as of the two sides of the flap. To alter the colour scheme, simply fold over the flap or turn the bag around. A modern bag for our times, for use in many different situations, with many variations, which satisfies the desire for change and for expressing one's personal lifestyle. The bag stands out by virtue of its innovative design, its many practical details and high-quality materials. Two large main compartments, clever partitioning on the inside and many practical pockets for small items provide room to stow away everyday items.

Jury ***Judges panel***

Eine Tasche mit vier Gesichtern, abhängig nach Tagesstimmung einsetzbar. Hohe Funktionalität, großer Stauraum, verschiedene Stautaschen, leichtes Material mit witzigen Spielmöglichkeiten. Wenn alle Zipper laufen ein perfektes Gerät. Die Zipper sind farblich sehr gut abgesetzt, so daß man auch alle Taschen findet. Sehr schön diese grafische Anordnung.

A bag with four faces, to be used depending on present mood. Very functional, plenty of storage room, different pockets, light material with amusing possibilities for play. As long as all the zips work, a perfect item. The zips' colour contrasts very well, allowing all the pockets to be found. This graphic arrangement is very attractive.

Hersteller ***Manufacturer***	**Design** ***Designer***
Bree Collection GmbH & Co. KG Gerberstraße 3 D-30916 Isernhagen Tel. +49 51 36 8 97 60 www.bree.de	Daniela Breuer, Christian Reichert Werksdesign

Flaschentasche „Conception 4" *bottle bag*

Fur die Entwicklung der Serie „Conception", zu der neben Taschen und Geldbörse diese Flaschentasche gehört, stand eine Papiertragetasche Pate. So sind die Taschen nach dem Prinzip der Falttechnik einer Papiertragetasche aufgebaut, eine einfache Idee mit einer bestechenden Ästhetik. Umgesetzt wurde es mit einem Materialmix aus strapazierfähigem Industriefilz als Trägermaterial und hochwertigem, pflanzlich gegerbten Rindleder. Auf dem Filz sind in einem kleinen Abstand zueinander Lederfelder angebracht, so daß dazwischen Filzfugen entstehen. Das ergibt ein interessantes Material- und Farbspiel. Die Flaschentasche zeichnet sich durch ihren geradlinigen Aufbau aus, der Modernität widerspiegelt.
Es gibt sie in den Farbkombinationen natur-schwarz, weiß-schwarz, schwarz-orange, beige-türkis und hellblau-brombeer. Dank des flexiblen Filzes läßt sich die extrem leichte, strapazierfähige Flaschentasche sehr klein zusammenfalten und so leicht aufbewahren, da sie wenig Stauraum benötigt.

The inspiration for the "Conception" series, which includes this bottle bag as well as handbags and a purse, was provided by a paper carrier bag. The bags have been constructed following the folding principles used for a paper carrier bag – a simple idea that is aesthetically very attractive. The idea has been put into practice with a mixture of hard-wearing industrial-grade felt, as backing material, and high-quality, naturally tanned cowhide. Leather patches are attached to the felt at short intervals, giving rise to felt seams in the interstices. The result is an interesting play of materials and colour. The bottle bag is linear in construction, giving it an air of modernity. It is available in the colour combinations natural-black, white-black, black-orange, beige-turquoise and sky blue-blackberry. The flexible felt allows this extremely light and very hard-wearing bottle bag to be folded up very small, making it easy to store away, as it takes up very little space.

Jury ***Judges panel***

Ein gut gestaltetes, witziges Produkt.

A well designed, witty product.

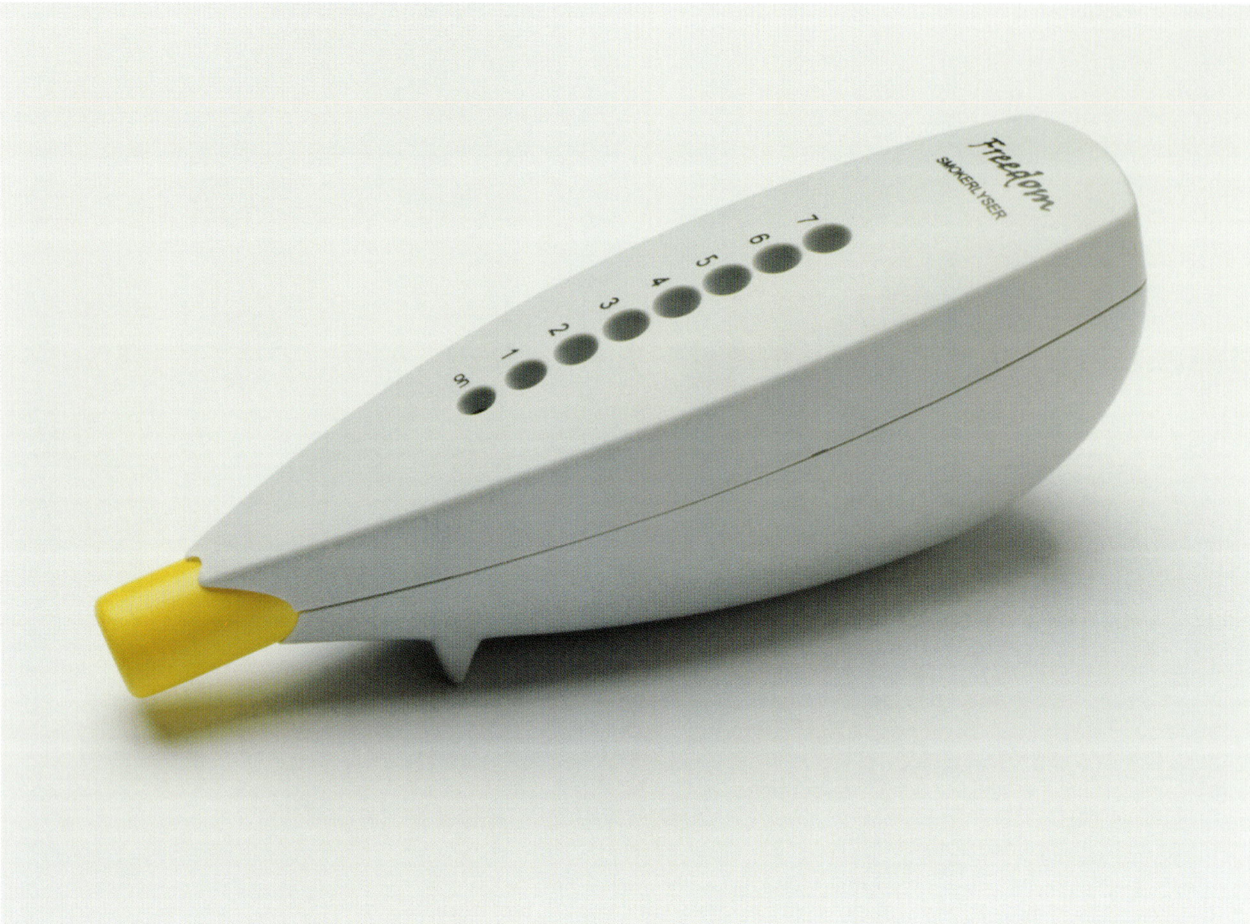
Freedom
SMOKERLYSER
on
1
2
3
4
5
6
7

Hersteller *Manufacturer*	**Vertrieb** *Distributor*	**Design** *Designer*
Bedfont LTD. 105 Laker Road, Rochester Airport ind. Estate Rochester, Kent ME1 3QX Tel. +44 16 34 67 37 20 www.bedfont.com	Neomed GmbH Danzigerstraße 93 D-41352 Korschenbroich Tel. +49 21 61 64 42 19 Fax +49 21 61 64 47 46	Uno Product Development Uri Friedländer 20 Belsize Avenue London, NW3 4AU Tel. +44 20 77 94 40 80

Raucheranalysegerät „Freedom Smokerlyzer®" *CO breath monitor*

Seit Mitte der achtziger Jahre werden weltweit in Raucher-Entwöhnungs-Seminaren erfolgreich Meßgeräte mit dem Namen „Smokerlyzer®" eingesetzt. Diese Geräte messen in der Ausatemluft den Anteil des sehr giftigen Gases Kohlenmonoxid (CO), das unter normalen Umständen nur bei Rauchern festzustellen ist. Stellt ein Raucher den Tabakkonsum ein, sinkt in wenigen Tagen der CO-Anteil auf das Niveau eines Nichtrauchers. Mit „Freedom" wurde nun erstmalig ein Gerät für den Privatgebrauch entwickelt, das der Betroffene zu Hause anwendet. Damit kann der entwöhnungswillige Raucher täglich seinen CO-Gehalt messen und den Erfolg sehen, was seine Willenskraft stärkt und ihm beim Durchhalten hilft. Ein Raucher, der 30 Zigaretten pro Tag konsumiert, kommt in der Meßskala auf den Wert sieben. Reduziert er seinen Konsum, sinkt das Kohlenmonoxidniveau, bis es den Wert eins erreicht. Das Gerät ist sehr nutzerorientiert gestaltet und stellt in der Formsprache und in der Funktion einen Bezug zum Rauchen her. Es ist leicht zu bedienen und liegt angenehm in der Hand. An der einfach und übersichtlich gestalteten Meßskala lassen sich die Werte gut ablesen.

At smoking cessation seminars throughout the world, measuring instruments known as "Smokerlyzers®" have been used successfully since the mid-1980s. These instruments measure the level of carbon monoxide (CO), which is a very poisonous gas, in end expired air. In normal circumstances, this gas will only be found in smokers. If a smoker quits consuming tobacco, his CO level will drop to that of a non-smoker within a few days. "Freedom" is now the first such instrument designed for private use, which the smoker can use in his own home. A smoker wanting to quit can measure his CO level daily and see the benefits of stopping smoking, and this in turn will strengthen his resolve and help him to persist in giving up. A 30-a-day smoker will have a value of seven on the measuring scale. If he cuts back his consumption, his CO level will drop until it reaches a value of one. The instrument has been designed in a very user-friendly way, and its formal expression and function allude to smoking. It is easy to operate and sits pleasantly in the hand. Values are easy to read on a simple, clearly arranged measuring scale.

Jury *Judges panel*

Interessantes Gerät, um Menschen vom Rauchen abzubringen. Die Belastungsanzeige ist dementsprechend. Wenn ich mit Rauchen aufhöre, kann ich selbst kontrollieren, wie weit meine Lunge von Schadstoffen befreit wird. Sicher ein gutes Hilfsmittel, um sich vom Rauchen zu befreien. Ein erzieherisches Gerät.

Interesting instrument to persuade people to give up smoking, with a measuring scale to match. If I stop smoking, I can see for myself how far my lungs have been cleared of pernicious substances. Without doubt, a good aid to ridding oneself of the smoking habit. An educational instrument.

Focus in Gold

Hersteller ***Manufacturer***	**Vertrieb** ***Distributor***	**Design** ***Designer***
Tonon & C. S.p.A. Via Diaz 22 I-33044 Manzano Tel. +39 04 32 74 07 40 Fax +39 04 32 74 07 70	WK Wohnen GmbH Im Gefierth 9a D-63303 Dreieich Tel. +49 61 03 39 10	Martin Ballendat Maximilianstraße 4 D-84359 Simbach am Inn Tel. +49 85 71 60 56 60

Armlehnstuhl/Loungechair „DE 773" *armchair/lounge chair*

Bei dem Stuhl sticht das ungewöhnliche schwarze Stricknetz als Komplettbezug von Rücken und Armlehnen ins Auge. Diese Stricktechnik kommt aus dem Kleidungsbereich. In verblüffend einfacher Form wird der gestrickte Bezug über einen durchlaufenden Rückenrohrrahmen gezogen. Durch unterschiedlich dichte Strickkonfiguration entstehen die dafür notwendigen Kedersäume und die eigentliche patentierte Innovation, ein miteingestrickter, gestalterisch prägnanter Gurt. Die Rückenlehnenfläche gibt dadurch beim Anlehnen leicht nach und ist so angenehm bequem und weich. Die luftige offene Netzstruktur läßt die Sitzmöbel leicht und transparent wirken. Der gestalterische Aufbau der Sessel ist aufgrund der geringen Komplexität der Bauelemente reduziert und zeitlos. Die Schalenoberteile aus Polyurethanschaum sitzen auf einem drehbaren Vierfußgestell aus Stahlrohr Chrom glänzend oder Chrom matt. Der höhere, um 180 Grad drehbare Armlehnstuhl läßt sich gut am Tisch einsetzen. Der größere und niedrigere Loungechair, der sich um 360 Grad drehen läßt, ist als Solitär etwa zur Polstergruppe gedacht. Die Sitzkissen sind in verschiedenen Farben und Dessins erhältlich.

The striking feature of this chair is the unusual black knitted webbing covering the entire back and arms. This knitting technique comes from the clothing sphere. In a surprisingly simple way, the knitted covering is pulled over an unbroken tubular seat-back frame. Varying the density of the knitting gives rise to the piping seams that are needed, as well as to the innovation that has actually been patented here, a belt that has been included in the knitting pattern. The surface of the seat back gives slightly when leant against, making it pleasantly comfortable and soft. The airy, open fishnet structure makes the seating appear light and transparent. The elements making up the chair are uncomplicated, making its design simple and timeless. The seat surfaces of polyurethane foam rest on a four-legged swivel frame of polished or matt chromium-plated tubular steel. The higher armchair, which can turn 180 degrees, is suitable for use at a table or desk. The larger and lower lounge chair, which can turn 360 degrees, has been conceived as a stand-alone chair to accompany a three-piece suite, for example. The cushions are available in various colours and designs.

Jury ***Judges panel***

Bestechend durch seine Zartheit und Anmut. Sehr schlanker, leichter, fast schwebender Stuhl, äußerst bescheiden in der Dimensionierung mit dünnen Sitzen und Fußgestellen, spielt eigenwillig mit Materialien, gute Rückenstützfunktion. Der erste transparente Polstersessel. Faszinierend. Eine neue, nicht eine klassische Eleganz. Eine archetypische Form.

Its delicacy and grace are captivating. Very slim, light, almost floating chair, extremely modestly dimensioned with thin seats and leg frames, plays unconventionally with materials, good back-supporting function. The first transparent upholstered chair. Fascinating. Its elegance is new, not classical. An archetypical form.

Hersteller *Manufacturer*	**Vertrieb** *Distributor*	**Design** *Designer*
Sedus Stoll AG Brückenstraße 15 D-79761 Waldshut Tel. +49 77 51 8 42 47 www.sedus.de	Sedus Stoll AG Brückenstraße 15 D-79761 Waldshut Tel. +49 77 51 8 42 47 www.sedus.de	Mathias Seiler Kirchstraße 17 D-79539 Lörrach Tel. +49 76 21 57 77 29 MathiasSeiler@sedus.de

Bürodrehstuhl „Sedus open up" *office swivel chair*

„Sedus open up" – der Stuhl zum Sitzen und Liegen. Am Schreibtisch nur aufrecht zu sitzen wird durch die moderne Informations- und Kommunikationstechnologie zunehmend in Frage gestellt. Dank einer neuen Synchronmechanik in Kombination mit einer innovativen Gestellstruktur neigt sich die Rückenlehne liegeähnlich. So bildet der Rückenlehnenträger ein Gelenk aus, an welchem der Rückenlehnenrahmen dorsokinetisch beweglich – das heißt, der Stuhlrücken läßt sich in jede Richtung frei bewegen – aufgehängt ist. Damit paßt sich der Stuhlrücken perfekt allen Bewegungen seines Nutzers an. Die Nackenstütze ist in Höhe und Neigung, die Armlehnen in Höhe und Tiefe individuell verstellbar, so daß Arme und Nacken auch in der Liegeposition wirksam unterstützt werden. Es wird nicht zwischen tragender und umhüllender Struktur unterschieden, die komplette Struktur ist ablesbar. Jedes Element ist als Sichtteil gestaltet. So läßt die aufgespannte, transparente Membran auch in der Vorderansicht die tragende Struktur der Rückenlehne erkennen. Die Formgebung des Rückenlehnenrahmens, die an ein Knochenskelett erinnert, visualisiert den Kräftefluß.

"Sedus open up" – the chair to sit and lie on. Modern information and communications technology is increasingly casting doubt on the idea that a desk is only there to be sat up straight at. Thanks to the new Synchron mechanism combined with an innovative frame structure, the back of the chair tilts like a reclining chair. The support for the back forms a joint on which the back frame is suspended "dorsokinetically" – in other words, the chair back can move freely in any direction. This allows the back of the chair to adapt perfectly to the user's every movement. Height and inclination of the neck support, as well as height and depth of the arm rests, can be adjusted individually, providing arms and neck with effective support in a reclining position, too. No distinction is made between load-bearing and surrounding structure, the entire structure is readable. Each element has been designed as a part to be seen. Even from the front, therefore, the transparent membrane covering the chair allows the supporting structure of the chair back to be seen. The design of the chair back frame, which is reminiscent of a skeleton, visualizes the flow of forces.

Jury ***Judges panel***

Weitere Reduzierung im Materialeinsatz mit sehr guter Funktion. Sitzen ohne Führerschein. Alemannische Akribie aber im Positiven, detailbesessen geradezu. Leichtigkeit in der Gesamterscheinung.

Further reduction in the use of materials, with very good function. Sitting without a driving licence. Alemannic attention to detail, but in a positive sense. Almost obsessed with detail. Overall impression of lightness.

Hersteller ***Manufacturer***	**Design** ***Designer***
Klöber GmbH	Starczewski Design Team
Rauensteinstraße 18	Heimstraße 29
D-88662 Überlingen	D-89073 Ulm
Tel. +49 75 51 83 81 46	Tel. +49 7 31 2 80 46
www.kloeber.com	Fax +49 7 31 2 73 80

Objektstuhl „Thira" *contract chair*

Der Objektstuhl „Thira" sticht durch seine Leichtigkeit und Schlankheit sowie durch Transparenz hervor. Dank der guten Funktion und dem hochwertigen schlichten Design läßt er sich vielseitig einsetzen: sowohl in repräsentativen Konferenz-, Besprechungs- und Seminarräumen als auch im Wohnbereich, etwa im Home-Office oder am Eßplatz. Der geradlinige Sitz scheint nahezu schwerelos im Vierbeingestell zu schweben. Eine flach gehaltene Polsterauflage, wahlweise in Leder oder Stoff, unterstreicht die schlanken Proportionen. Für guten Sitzkomfort sorgen zusätzlich die in die Sitzfläche gespannten Gummigurte. Das formal ganz reduzierte Gestell ist aus Rundrohr verchromt, schwarz oder aluminiumfarbig hergestellt. Der Sitz besteht aus einer Sperrholz-Formholzschale und ist in Buche furniert, natur oder allen Beiztönen zu haben. Die zierlichen Armauflagen gibt es entweder in Buche massiv oder mit Leder bezogen.

The "Thira" contract chair is striking for its lightness and slimness, as well as its transparency. Because it is so functional and thanks to its high-quality, unassuming design, it can be used in many ways: both in conference, meeting and seminar rooms where appearance is important, and in the home – in the home-based office, for example, or at the dining table. The straight lines of the seat seem to float almost weightlessly on the four-legged frame. The seat's slimness is underlined by the upholstered cushion, which has been kept flat and is available in leather or fabric. Further seating comfort is achieved by the rubber belts stretched across the seating area. The formally very reduced frame is made of chromium-plated tubular steel, available in black or aluminium. The seat is made of a plywood shell, and is available with a beech veneer, untreated or stained in any colour. The delicate top surfaces of the arm rests are available in solid beech or covered with leather.

Jury ***Judges panel***

Moderner Klassiker, hoher Sitzkomfort, sehr einfach gelöst, aber doch wiedererkennbare Gestaltprinzipien durch Anbinden von Rücken und Beinen an Armlehne. Schöner Stuhl. Drei Materialien, die sehr gut harmonieren.

Modern classic, very comfortable to sit on, very simply solved, yet recognizable design principles in the way the back and legs have been attached to the armrests. Beautiful chair. Three materials that harmonize very well.

Hersteller ***Manufacturer***	**Vertrieb** ***Distributor***	**Design** ***Designer***	
Performa Möbel und Design GmbH Marbacher Straße 54 D-74385 Pleidelsheim Tel. +49 71 44 89 88 69 www.performa.de	Performa Möbel und Design GmbH Marbacher Straße 54 D-74385 Pleidelsheim Tel. +49 71 44 89 88 69 www.performa.de	a/2 Thomas Albrecht Kattenstraße 8 D-34119 Kassel Tel. +49 5 61 7 66 76 18	Martin Kleinhans Werksdesign

Tisch „Metatisch" *table/desk*

„Metatisch" ist die einfachste Form eines Tisches: Tischplatte und vier Beine. Er ist auf das Wesentlichste reduziert und hat überall dort seinen Platz, wo Modularität und Geradlinigkeit gefragt ist. Ursprünglich wurde er als Bürotisch konzipiert, an den alle weiteren Funktionselemente nach individuellem Bedarf modular dazuaddiert werden können. Durch intelligente Details wird der Tisch vielseitig verwendbar. So sind die Tischbeine individuell in der Höhe von 71 auf 81 Zentimeter zu verstellen. Mit einem einfachen Werkzeug kann der Tisch, wird er gerade mal nicht gebraucht, zerlegt werden. Die Tischplatte läßt sich, da sie ein planes Element ist, ideal stapeln und platzsparend verstauen. „Metatisch" ist ein Mehrzwecktisch, der viele Gestaltungsfreiräume bietet. So kann er als Eß-, Büro- und Familientisch eingesetzt werden. Die Tischplatte besteht aus Schichtholz mit Linoleumfläche, die eine fühlbar angenehme Haptik hat und in 16 verschiedenen Farben zu haben ist: von neutraler zurückhaltender Farbigkeit bis zur akzentuierten Farbfläche. Das Gestell ist aus pulverbeschichtetem Stahl. Der Tisch wird mit CAM (Computer Aided Manufacturing) zeitgemäß und preisgünstig gefertigt.

"Metatisch" is a table in its simplest form: table top and four legs. It has been reduced to the bare essentials, and is at home wherever a modular, no-frills approach is wanted. It was originally designed as a desk to which other functional modules can be added according to individual requirements. Intelligently thought out details make the table very versatile. The legs can be adjusted individually to any height between 71 and 81 centimetres. If the table is not needed for a while, it can be taken apart with one simple tool. As the table top is flat, it is ideal for stacking and storing away in the smallest of spaces. "Metatisch" is a multi-purpose table that leaves its user with plenty of choice as to how to use it. It can be used as a dining or family table, or as a desk. The table top is made of plywood with a linoleum surface, which is pleasant to the touch and is available in 16 different colours: from subdued tones to a bold splash of colour. The frame is made of powder-coated steel. The table is manufactured in a modern and inexpensive way, using CAM.

Jury ***Judges panel***

Vier Beine braucht man, um eine Platte vom Boden abzuhalten. Das ist hier konsequent umgesetzt. Durch die Reduziertheit sehr charmant. Es ist gelungen, auf einfachste Weise, eine Materialkombination auch technisch zu bewältigen, das heißt die Leichtigkeit des Tisches und die Stabilität sind gleichermaßen gewährleistet. Schönes und dezentes Spiel mit Materialien.

All you need to hold a top above the floor is four legs. This has been put rigorously into practice here. Its reduction makes it charming. In the simplest way, a combination of materials has been achieved technically as well, with lightness and stability equally being guaranteed in the table. Beautiful, discreet play with materials.

Hersteller ***Manufacturer***

Leitner GmbH
Düsseldorfer Straße 14
D-71332 Waiblingen
Tel. +49 71 51 1 70 60
www.leitner.de

Vertrieb ***Distributor***

Leitner GmbH
Düsseldorfer Straße 14
D-71332 Waiblingen
Tel. +49 71 51 1 70 60
www.leitner.de

Design ***Designer***

Rolf Heide
Drosselweg 5
D-22926 Ahrensburg
Tel. +49 41 02 5 64 70

Tisch „Serie L+" *table/desk*

Der formal klar und reduziert gestaltete, hochwertige Tisch „Serie L+" kann je nach Nutzergruppe gesondert ausgestattet werden und erfüllt damit unterschiedlichste Funktionen. So macht ihn eine edelholzfurnierte, mit Linoleum oder Schichtstoff belegte Platte zum Schreibtisch. Mit Glasplatten wird er zum Präsentations- und, noch zusätzlich mit indirekter Beleuchtung ausgerüstet, zum Exponattisch. Wenn er mit segmentierten herausnehmbaren Platten ausgestattet ist, kann er sogar als Stauraumtisch eingesetzt werden. Sondermaße sind möglich.

With its formal clarity and reduced design, "Serie L+" is a high-quality table that can be varied according to user and hence perform the widest variety of functions. A top of high-grade wood veneer, linoleum or laminate, for example, makes it into a desk. If a glass top is chosen, it becomes a presentation table, or if additionally equipped with indirect lighting, a display table. If it is fitted with compartmented, removable leaves, it can even be used as a storage table. Also available in special sizes.

Jury ***Judges panel***

Ein schlichter und vielseitiger Präsentationstisch, ideal für den gehobenen Ladenbau, ermöglicht die unterschiedlichsten Lösungen der Präsentation und Beleuchtung und ist gleichzeitig als eigenständiges Gestaltungselement im Raum eine spannende Lösung. Mehrere Funktionen in einem.

A modest, versatile presentation table, ideal as up-market shop furniture, allowing many different ways of presenting and illuminating. At the same time, an intriguing piece of furniture as a design element in its own right. Several functions in one.

Hersteller ***Manufacturer***	**Vertrieb** ***Distributor***	**Design** ***Designer***	
Ceka Büromöbelwerke Erich-Krause-Straße 1 D-36304 Alsfeld Tel. +49 66 31 18 62 62 www.ceka.de	Ceka Büromöbelwerke Erich-Krause-Straße 1 D-36304 Alsfeld Tel. +49 66 31 18 60 www.ceka.de	B-Plan Rathlefstraße 7 D-31582 Nienburg/Weser Tel. +49 50 21 92 52 28	Ceka Werksdesign

Besprechungstisch „Smartform" *meeting table*

Bei dem modernen und ansprechenden Besprechungstisch verbindet sich Form mit Funktion und Technik zu einer harmonischen Einheit. Markantes Gestaltungselement ist das funktionelle Kapitell, das gleichzeitig perfekt das Thema Höhenverstellung löst. Durch leichtes Drehen läßt sich die gewünschte Arbeitshöhe von 66 bis 78 Zentimeter einstellen und individuell den Bedürfnissen des arbeitenden Menschen anpassen. Ein Funktionsmerkmal wird zum Designelement. Wer will, kann einen farbigen Blickfang schaffen und die Kunststoffabdeckungen am funktionellen Kapitell anstelle von Standard schwarz auch in blau, grün, orange oder weiß aussuchen. Die Arbeitsplatte ist ebenfalls individuell wählbar, so gibt es Platten in Melamin, Furnier und die filigrane Kompaktplatte HPL, die nur zehn Millimeter stark ist. Der Besprechungstisch gehört zu einem umfangreichen Büromöbelprogramm. Dazu zählen neben Arbeits-, Konferenz- und runden Besprechungstischen auch ein Steh- und ein Bistrotisch.

This modern and appealing meeting table is a harmonious combination of form, function and technology. Its most striking design element also performs a function: the capital, which presents a perfect solution to the problem of adjusting height. Simply by turning, the desired working height can be adjusted to any height between 66 and 78 centimetres, customizing it to the individual needs of the working person. A functional feature has become a design element. If so desired, the plastic cover for the capital can be highlighted by choosing it in blue, green, orange or white instead of the standard black. The work top can also be chosen as desired, and tops are available in melamine, veneer and the filigree HPL compact top, which is only ten millimetres thick. The meeting table is part of an extensive range of office furniture. Apart from desks, conference tables and round meeting tables, there is also a high desk and a bistro table.

Jury ***Judges panel***

Faszinierend einfache Höhenverstellung, die wirklich Sinn macht, mit einem fast schwebend anmutenden Tischkonzept, in dem die Arbeit zum Wesentlichen wird und nicht der Tisch.

Fascinatingly simple height adjustment that truly makes sense, together with a table design that almost appears to float, and where the main issue is work, not the desk.

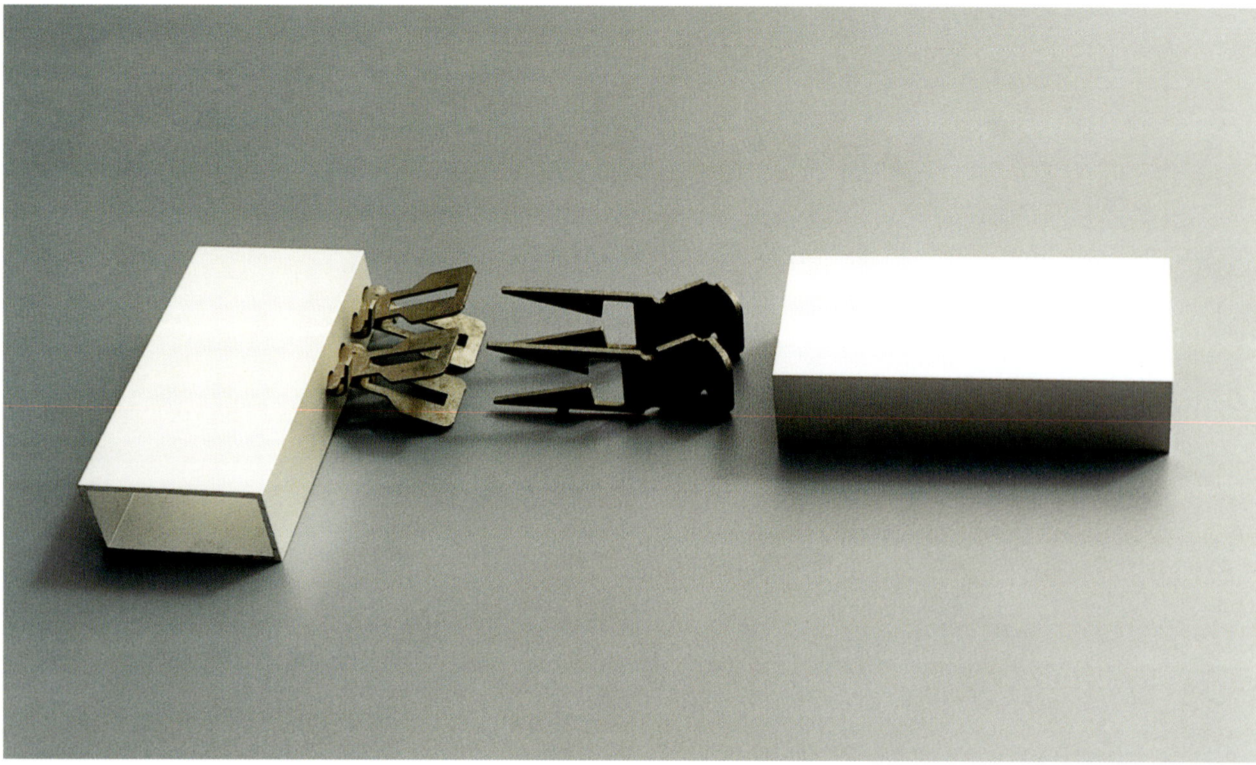

Hersteller ***Manufacturer***	**Vertrieb** ***Distributor***	**Design** ***Designer***
Dalli-Klick Josef Sekler Hasenbergstraße 24 D-73479 Ellwangen-Pfahlheim Tel. +49 79 65 80 06 32 www.wir-einrichtungen.de	Dalli-Klick Josef Sekler Hasenbergstraße 24 D-73479 Ellwangen-Pfahlheim Tel. +49 79 65 80 06 32 www.wir-einrichtungen.de	Josef Sekler Werksdesign

Rohrstecksystem „Dalli-Klick" *tube connecting system*

„Dalli-Klick" ist ein patentiertes Rohrverbindungssystem, das völlig im Innern der Rohre verschwindet. Das ergibt formschlüssige und von außen nicht sichtbare Verbindungen. Es lassen sich Rohrkonstruktionen erzeugen, die vom Möbel über Messestände bis zu räumlichen Strukturen reichen. Die Verbindung wird werkzeuglos ohne Schrauben zusammengesteckt und ist äußerst formstabil. Trotzdem läßt sie sich sehr einfach lösen, es genügt ein einfacher Hammerschlag. Die intelligente Fügetechnik ist Basis für eine enorme Variabilität des Systems. Es gehört zu einem Messe-, Möbel- und Ladenbausystem, das aus Rund- und Rechteckrohren aus Stahl, Edelstahl oder Aluminium besteht. Da das Rohrverbindungssystem unsichtbar ist, läßt es sich aber genausogut bei Rohren aus anderen Materialien, etwa Messing oder Chrom, einsetzen.

"Dalli-Klick" is a patented tube connecting system that disappears completely inside the tube. The result is a flush connection that is not visible from the outside. This allows tube constructions to be put together ranging from furniture via trade fair stands to spatial structures. The connection is plugged together without tools or screws, and is extremely stable. Nevertheless, it can be unplugged again very easily; a simple tap with a hammer is all that is needed. Intelligent joining technology gives the system its enormous versatility. It is part of a trade fair, furniture and shop-fitting system, which is made up of round or rectangular tubes made of steel, stainless steel or aluminium. As the tube connecting system is invisible it can equally well be used in tubes made of other materials such as brass or chrome.

Jury ***Judges panel***

Bestechend durch die Einfachheit. Man sieht keine Verbindungselemente. Man vermutet, es könnte geklebt sein, doch es handelt sich um eine einfache mechanische Verbindung, die genial gelöst wurde.

Captivatingly simple. No connecting elements can be seen. It gives the feeling that the tubes have been stuck together, yet it is a simple mechanical connection that has been solved brilliantly.

Hersteller ***Manufacturer***	**Vertrieb** ***Distributor***	**Design** ***Designer***
Formfarm Neuer Berg 24 D-70327 Stuttgart Tel. +49 7 11 3 04 00 70 www.formfarm.de	Formfarm Neuer Berg 24 D-70327 Stuttgart Tel. +49 7 11 3 04 00 70 www.formfarm.de	Benjamin Gugelberger, Ulf Braun Werksdesign

Regal und Möbelsystem „Spin off" *shelving and furniture system*

„Spin off" läßt sich werkzeuglos montieren, schnell umbauen und jederzeit leicht erweitern. Vergleichbare Möbelsysteme brauchen hierfür einen höheren Zeitaufwand, Spezialwerkzeug sowie externes Montagepersonal. Das Kernstück des Möbelsystems ist ein neues, revolutionäres, weltweit patentiertes Verbindungsprinzip. Dabei wird ein Steckbolzen aus glasfaserverstärktem Kunststoff mit einem patentierten Aluminium-Profilstab verpreßt. Profilstab und Verbinder werden einfach zusammengesteckt, von Hand leicht gedreht, so daß beide spielfrei und auszugsicher miteinander verbunden sind. So läßt sich im Handumdrehen die offene selbsttragende Grundstruktur werkzeuglos zusammensetzen. In diese werden anschließend Ablagen und Zwischenwände eingesetzt, die dank kleiner Prägungen an den Kanten spielfrei in die Profilnut einrasten. Mit einer zusätzlichen Stahlfeder werden stärker belastete Seitenwände, für Auszüge und Türen, gesichert. Dem Möbelsystem liegt ein gut durchdachtes Raster zugrunde. Die Funktionselemente wie Schubfächer, Regalablagen, Klappen und Hängeregister lassen sich ebenfalls werkzeuglos einsetzen.

"Spin off" can be assembled without tools, quickly re-arranged and easily extended at any time. These tasks take up a greater amount of time with comparable furniture systems, as well as requiring special tools and workers to put them together. The key element of the furniture system is a new, revolutionary, internationally patented connecting principle. During manufacture, a socket pin of glass fibre-reinforced plastic is pressed into a patented aluminium-section bar. The bar and a connecting element are simply clipped together and turned slightly by hand, leaving them connected free of play and firmly in place. The open, self-supporting base structure can therefore be put together in a trice, without any tools. Shelves and partitions can then be inserted into this structure, and thanks to their slightly embossed edges, the former clip into the section groove free of any play. An additional steel spring supports side walls bearing a greater load, such as drawers or doors. The furniture system is based on a logical pattern. Functional elements such as drawers, shelves, doors and filing cabinets can also be inserted without tools.

Jury ***Judges panel***

Einfaches Regalsystem, das keine Werkzeuge benötigt und selbsterklärend ist. Vielseitige Einsatzmöglichkeiten, doppelte Wandführung, dadurch die Möglichkeit, in Zwischenwänden weitere Präsentationsobjekte unterzubringen, ideal für Messe- und Ladenbau. Eine weitere Variante in der Vielzahl solcher Systeme. Bestechend jedoch der werkzeuglose Aufbau.

Simple shelving system that needs no tools and is self-explanatory. Many possibilities for use, double wall, thus allowing further presentation objects to be positioned in partition walls, ideal for trade fairs and shop furnishing. Another variant to the many such systems available. Stands out because it can be assembled without tools, however.

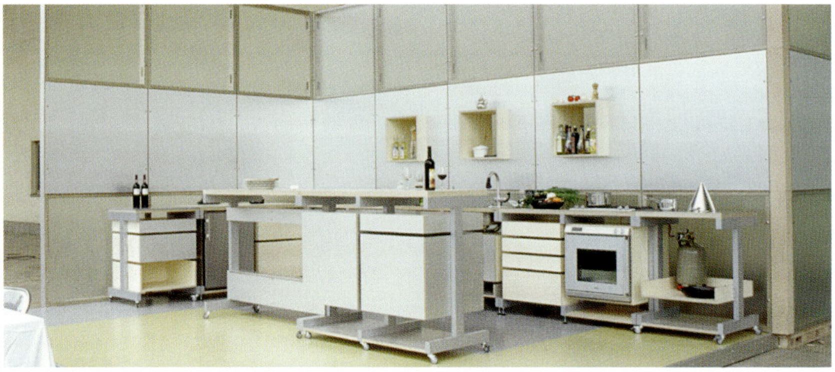

Hersteller ***Manufacturer***

Eberlein GmbH
Ansbacher Straße 126
D-90449 Nürnberg

Engelbreit & Sohn
Marco Engelbreit
Heinrich-Diehl-Straße 25
D-90552 Röthenbach

Vertrieb ***Distributor***

3L Modulbausystem
Jürgen Steinhauser
Steinstraße 21
D-90419 Nürnberg
Tel. +49 9 11 77 60 70
www.loftschau.de

Design ***Designer***

3L Modulbausystem
Jürgen Steinhauser
Steinstraße 21
D-90419 Nürnberg
Tel. +49 9 11 77 60 70
www.loftschau.de

System „3L Modulbausystem" *system*

Das System reagiert schnell auf neue Bedürfnisse vom mobilen Büro bis zum komplett ausgestatteten Großraumbüro. Das Basismodul ist ein Quadrat aus Holz. Dank des Bausteinsystems sind die Elemente schnell zu zerlegen, leicht zu transportieren und neu anzuordnen. „3 L" ist ein offenes System. Es ist in sich unabhängig und gleichzeitig völlig frei in seinen Kombinationsmöglichkeiten. So kann es beispielsweise beidseitig mit Holz beplankt werden, will man blickdichte Wände. Oder es lassen sich satiniertes Glas bis zu lichtvollen, transparenten Wänden aus Glas und Plexiglas einsetzen, was den Arbeitsraum kommunikativ und offen werden läßt. Beplankt man nur einseitig, können die einzelnen Segmente des Systems gleichzeitig als Regal benutzt oder, mit Leuchtmodulen ausgestattet, lichttechnisch gestaltet werden. Es läßt sich auch mit einem Küchenmodulsystem kombinieren. Dadurch können Küchenzeilen und Thekeninstallationen gestaltet werden, die feststehend oder rollbar sind und vielseitig eingesetzt werden können: zu Hause, auf einer Messe, bei Dreharbeiten oder für besondere Anlässe. Dabei wird eine Rundum-Dienstleistung bis hin zum Catering angeboten.

This system reacts rapidly to the new needs of mobile office right up to the completely equipped open-plan office. The basic module is a wooden square. Thanks to the modular system, the elements can quickly be knocked down, easily transported and re-arranged again. "3 L" is an open system. It is independent and allows unlimited possibilities for combination. Wooden panels can be attached to both sides, for example, if opaque walls are needed. Translucent glass can also be inserted, or even walls of glass or Plexiglass that allow light to shine through completely, making the office communicative and open. If only one side is boarded up, the individual segments of the system can also be used as a shelf, or be used as a basis for lighting features. It can also be combined with a modular kitchen system. This allows kitchen ranges and counters to be designed that are either fixed or mobile, and can be used in many different ways: at home, at a trade fair, when shooting a film, or for special occasions. An all-round service is provided, right up to catering.

Jury ***Judges panel***

Radikale Architektur- und Möbelkombination, dazu eine sehr einfache Lösung für die Integration von Küchenmöblierungen gekoppelt mit interessantem Serviceaspekt. Dennoch eine gewagte und anspruchsvolle Lösung für Messe- und Eventbau, mobil, stringent und konsequent durchdacht. Im Detail sollten noch gestalterische Qualitäten herausgearbeitet werden.

Radical combination of architecture and furniture, together with a very simple solution for the integration of kitchen furniture, coupled with an interesting service aspect. Nevertheless, a daring and sophisticated solution for trade fair and event stand systems, mobile, consistently and rigorously thought through. Some design qualities could still be brought out in the smaller areas of detail.

VIEW

Hersteller ***Manufacturer***	**Vertrieb** ***Distributor***	**Design** ***Designer***
Reisenthel Accessoires Benzstraße 3 D-82178 Puchheim Tel. +49 89 8 00 64 40 Fax +49 89 80 60 35	Reisenthel Accessoires Benzstraße 3 D-82178 Puchheim Tel. +49 89 8 00 64 40 Fax +49 89 80 60 35	Katja Horst Amalienstraße 15 D-80333 München Tel. +49 89 28 75 51 31 mail@KatjaHorst.de

Paravent „Claro" *screen*

Der Paravent zeichnet sich durch Einfachheit und Klarheit aus. Mit ihm können sowohl im Büro als auch zu Hause verschiedene Bereiche abgetrennt, Dinge versteckt, neue Raumeinheiten geschaffen und immer wieder verändert werden. Die Wände sind lichtdurchlässig und leicht transparent. Es gibt eine Single- und eine Duo-Ausführung, die sich bei Bedarf mit Anbauelementen erweitern lassen. Das multifunktionale patentierte Stecksystem läßt sich stufenlos und kinderleicht in verschiedenen Winkeln aufstellen. Die kegelförmigen, gußeisernen Füße geben dem Paravent einen festen Stand. Die Wände sind aus dem sehr strapazierfähigen Polypropylen, das später recycelt oder thermisch verwertet werden kann. Die Längsprofile wurden aus Aluminium gefertigt. Es gibt verschiedene Varianten: von der neutralen Wand „Plain Milky" bis „Pockets", wo sich in Schnitte mehrere Taschen einhängen lassen. Die Ausführung „Curly" hat schneckenförmige Schnitte, in die Zettel, Karten, Fotos und anderes gesteckt werden können. Weitere Versionen wurden mit dekorativen Siebdrucken mit Motiven wie „Orchidee" und „Palme" versehen.

The distinguishing features of this screen are its simplicity and clarity. Both in the office and at home, it can be used to separate off various areas, to conceal things and to create new room units, and be re-arranged again and again. The walls are translucent and slightly transparent. There is a "single" and a "duo" version, both of which can be extended if necessary with extension elements. The multifunctional, patented clip system can be erected extremely simply and at any angle. The conical, cast iron feet provide the screen with a sturdy support. The walls of the screen are made of very hard-wearing polypropylene, which can later be recycled or burned. The vertical sections are made of aluminium. Various versions are available, from the neutral "Plain Milky" screen to "Pockets", which is cut to allow several pockets to be inserted. The "Curly" version is cut spirally to allow pieces of paper, cards, photographs and the like to be held in place. Other versions have been decorated with screen prints featuring subjects such as "orchid" and "palm".

Jury ***Judges panel***

Völlige Vereinfachung des Prinzips Paravent, transparent, leicht, unaufdringlich. Einfach und elegant, vielfältig einsetzbar. Hohe Mobilität mit zwei adäquaten Materialien.

Absolute simplification of the screen principle, transparent, light, unassuming. Simple and elegant, for use in a variety of situations. Very easy to move, using two suitable materials.

Eskimo

VOGUE
VIEW ON COLOUR
13
THE COLOUR FORECASTING BOOK
VOGUE
INTERIOR VIEW 14
blind design
VOGUE
PELLE
INTERIOR VIEW 11
on the edge
PELLE
INview 16
GOLD
INTERIOR VIEW 12
composition

Hersteller ***Manufacturer***

Reisenthel Accessoires
Benzstraße 3
D-82178 Puchheim
Tel. +49 89 8 00 64 40
Fax +49 89 80 60 35

Vertrieb ***Distributor***

Reisenthel Accessoires
Benzstraße 3
D-82178 Puchheim
Tel. +49 89 8 00 64 40
Fax +49 89 80 60 35

Design ***Designer***

Katja Horst
Amalienstraße 15
D-80333 München
Tel. +49 89 28 75 51 31
mail@KatjaHorst.de

Zeitschriftenhalter „Redondo" *magazine rack*

„Redondo" sorgt für Ordnung unter Zeitschriften, Magazinen sowie Prospekten und wird dank seinem Objektcharakter zum Blickfang in Büro, Entree, Wartezonen, Verkaufsräumen, auf Messeständen wie auch im Wohnbereich. Im Sinne eines emotionalen Funktionalismus vereint der Zeitschriftenhalter modernes Design und hohe Funktionalität. Semitransparente Fächer sorgen dafür, daß man den Durchblick über das Mediensortiment behält. Sie bieten jeweils Platz für bis zu vier Zeitschriften. Der leichte Gegendruck des darüberliegenden Fachs sorgt dafür, daß sämtliche Printmedien stramm gehalten werden. Die Gestaltung ist auf der einen Seite pur und minimalistisch kühl, auf der anderen Seite weich und sinnlich ansprechend. Ein weiterer spannender Kontrast wird durch den Materialmix von kühlem Aluminium mit dem leichten, haltbaren und zugleich biegsamen Polypropylen erzeugt. Beide hochqualitative Materialien sind perfekt verarbeitet und entsprechen dem Mono-Materialkonzept. Außer dem Zeitschriftenhalter mit zehn Fächern gibt es noch zwei andere Modelle einmal mit sieben und mit vier Fächern. Die Fächer sind in den Farben milky und smoke erhältlich.

"Redondo" keeps newspapers, magazines and brochures neat and tidy. Its sculptural qualities mean that it will be an eye-catcher in the office, the lobby, in waiting areas, sales premises, in trade fair booths and in the home. True to the tenets of emotional functionalism, the magazine rack unites modern design and a high level of functionality. Semitransparent pockets ensure that you do not lose track of all the media kept inside. Each pocket has space for up to four magazines. Slight pressure from the pocket above ensures that all the printed media are kept firmly in their place. The design is pure and minimalistically cool on the one hand and soft and sensually appealing on the other. A further fascinating contrast is generated by its mix of materials, with cool aluminium and light, durable and yet flexible polypropylene. Both high-quality materials have been perfectly finished and satisfy the principle of recyclability. Apart from the magazine rack with ten pockets, there are two further models with seven and four pockets. The pockets are available in the colours milky and smoke.

Jury ***Judges panel***

Verblüffend einfache formale Lösung. Raumsparend und kompakt. Spannende Formgebung durch das gezogene, gerundete Profil. Wirkt fast wie eine Brotdose. Beeindruckend die Konstruktion, mit welch einfachen Mitteln Extrudermaterial im Kunststoff- und Aluminiumbereich kombiniert wird. Sehr ästhetisches und gefälliges Produkt, leicht und grazil.

Surprisingly simple formal solution. Space-saving and compact. The drawn, rounded profile makes its form very exciting. It almost looks like a kind of sandwich tin. Its design is impressive, using quite simple means to combine extruded plastic and aluminium. An extremely aesthetic, pleasing product that is light and graceful.

STABILO

Hersteller ***Manufacturer***	**Vertrieb** ***Distributor***	**Design** ***Designer***
Schwan-Stabilo Schwanhäußer GmbH & Co. Schwanweg 1 D-90562 Heroldsberg Tel. +49 9 11 5 67 13 65 www.stabilo.com	Schwan-Stabilo Marketing Deutschland GmbH & Co. Schwanweg 1 D-90562 Heroldsberg Tel. +49 9 11 5 67 32 66 www.stabilo.com	Factor Product Boris Simon, Frank Thiele Comeniusstraße 1 D-81667 München Tel. +49 89 48 92 78 10

Gelroller „STABILO® 's move® the gelball" *gelball pen*

Der Gelroller ist Mitglied der Produktfamilie „Stabilo 's move" und erfüllt damit alle Merkmale, die ein 's move haben muß: Spitze, Griff und Schreibgefühl – alles außergewöhnlich. Seine fast unverwüstliche, rollende Spitze bringt das Schreibgel konstant und zuverlässig aufs Papier. Der sehr griffige Schaft besitzt transparente Streifen, durch die man die noch verbleibende Gelmenge kontrollieren kann. Dank der typischen 's move Form und seiner Gel-Technik bietet der Stift ein ganz neues, gleitendes Schreibgefühl. Darüber hinaus ist er der erste 's move, der nachfüllbar ist. Durch sein Refill-System läßt sich die Gel-Tinte schnell und leicht wieder auffüllen. Ermöglicht wird dies durch den völlig neuen Bajonett-Verschluß. Dieser wird durch kurzes Drücken und Drehen mit aufgesetzter Kappe entriegelt. Eine weitere Neuheit sind die Metallic-Farben, die sogar auf dunklem Papier schreiben: gold metallic, blau metallic und dunkelgrün metallic. Die Tinte gibt es aber auch in gewohntem blau und schwarz.

Because this gel roller is a member of the "Stabilo 's move" product family, it lives up to all the qualities a 's move must have: point, barrel, and the way it feels when writing – all of them are extraordinary. Its almost indestructible, rolling point ensures that the flow of gel to the paper is constant and reliable. Transparent strips have been inserted in the extremely easy to grip barrel, allowing the level of gel remaining to be checked. Thanks to the typical 's move form and its gel technology, the pen offers a completely new, gliding writing experience. Moreover, it is the first 's move that can be refilled. Its refill system allows the gel ink.to be replenished quickly and easily. This is made possible by a completely new bayonet catch, which is released when the cap is on by a short press and turn movement. The metallic colours – gold, blue and dark-green – are a further new feature. They will even write on dark paper. But ink is also available in the usual blue and black.

Jury ***Judges panel***

Angenehme Haptik, leichte Führung der Spitze, gleichzeitig ein peppiges und sportliches Design. Animiert zum Kauen. Damit kann man sich das Zahnfleisch massieren. Bei diesem Schreiber ist die allgemeine Tendenz sehr deutlich, daß man Materialhaftigkeit über das haptische Moment erlebt.

Pleasant to the touch, point is easy to guide, a design that is both up-beat and sporty. Incites you to chew. You can massage your gums with it. This pen very clearly illustrates the general trend towards using material properties to make holding pleasant.

STABILO BOSS
MINI
STABILO BOSS
MINI
STABILO BOSS
MINI
STABILO BOSS
MINI
STABILO BOSS
MINI

Hersteller ***Manufacturer***	**Vertrieb** ***Distributor***	**Design** ***Designer***
Schwan-Stabilo Schwanhäußer GmbH & Co. Schwanweg 1 D-90562 Heroldsberg Tel. +49 9 11 5 67 13 65 www.stabilo.com	Schwan-Stabilo Marketing Deutschland GmbH & Co. Schwanweg 1 D-90562 Heroldsberg Tel. +49 9 11 5 67 32 66 www.stabilo.com	Edward Winkelbauer Industrial Design Myliusstraße 3 D-71638 Ludwigsburg Tel. +49 71 41 90 32 22

Highlighter „STABILO BOSS® MINI" *marker pen*

Der Klassiker „Stabilo Boss" wurde nun auch im Mini-Format herausgebracht. Der Mini verfügt dabei über die gleichen Merkmale wie der große Bruder: pigmentierte Tinte, Strichstärke und fünf verschiedene Farben. Er ist jedoch ein eigenständiges Produkt mit einer eigenen, jungen Zielgruppe. So sehen laut einer Studie die 13- bis 17-Jährigen den Mini gerade wegen seiner witzigen Proportionen und angenehmen Form nicht nur als Schreibutensil, sondern auch als Spaßprodukt. Da er gut in Hand-, Jacken- oder Hosentaschen paßt, läßt er sich überallhin mitnehmen. Außerdem ist der Mini in witzigen Verpackungseinheiten, wie etwa drei oder fünf verschiedenen Farben jeweils in einer Tüte, zu haben.

A miniature version of the classic "Stabilo Boss" has now been brought out. The Mini has the same qualities as its big brother: pigmented ink, width of tip, and five different colours. But this is a product in its own right, with a young target group of its own. According to a study, it is precisely because of the Mini's amusing proportions and pleasant form that 13 to 17 year olds regard it not only as a writing implement, but also as a fun product. As it fits easily into handbags, jacket pockets or trouser pockets, it can be taken everywhere. In addition, the Mini is available in fun-sized packs; for example, three or five different colours in one bag.

Jury ***Judges panel***

Faszinierend, daß man mit Stümperhaftigkeit im Sinne des Wortes einen Preis bekommen kann. Die Marketingexperten nennen ihn „stubsig". Er ist aber nicht nur niedlicher Handschmeichler, sondern auch funktional insofern, daß kleine Füllmengen das Austrocknen verhindern. In letzter Zeit ist ja auch das Relaunch von einem alten Produkt wieder im Kommen, wie etwa mit dem Käfer und anderen.

You too can be short and thickset – the marketing people would say "stubby" – and win a design prize. It isn't just a cute object that fits nicely into the hand, but functional too, since small amounts of ink are less likely to dry up. In recent years, we have increasingly seen people relaunching old products, as was the case with the Beetle and other products.

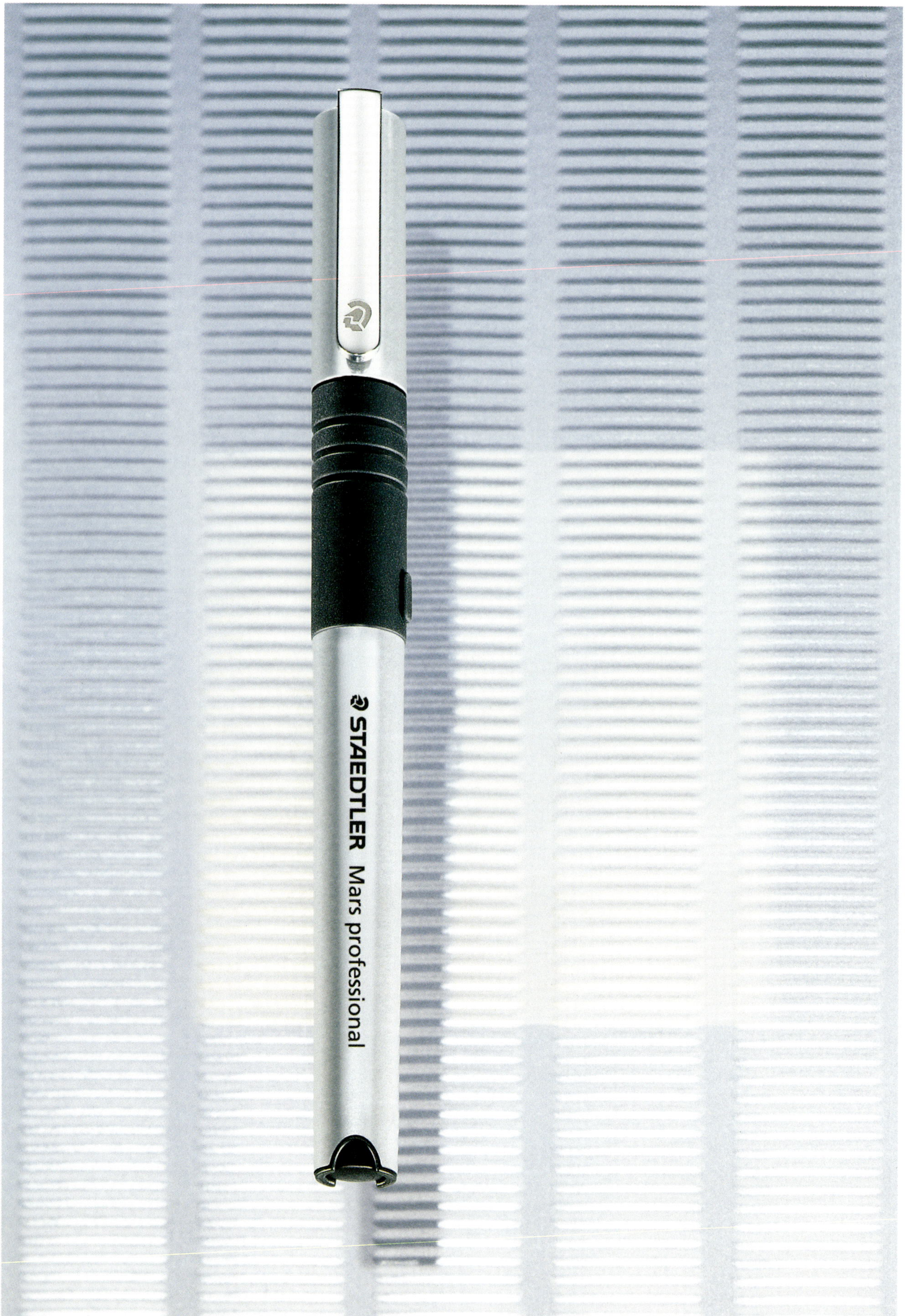
STAEDTLER
Mars professional

Hersteller ***Manufacturer***	**Vertrieb** ***Distributor***	**Design** ***Designer***
Staedtler Mars GmbH & Co Moosäckerstraße 3 D-90427 Nürnberg Tel. +49 9 11 9 36 50 www.staedtler.de	Staedtler Deutschland Vertriebs-GmbH Moosäckerstraße 3 D-90427 Nürnberg Tel. +49 9 11 9 36 52 32	Staedtler Design Team

Schreib- und Zeichenstift „STAEDTLER® Mars Professional"
writing and drawing pen

Der Stift ist ein innovatives Schreib- und Zeichengerät in höchster Präzision und komfortablem Handling. Er zeichnet sich neben funktionalem Design und hochwertiger Aluminium-Optik durch viele technische Raffinessen aus. So schützt die einzigartige korrosionsbeständige Metallfassung die Spitze und stabilisiert beim Schreiben und Zeichnen mit Linealen und Schablonen. Die Hightech-Kunststoffspitze garantiert zusammen mit speziell abgestimmten Tuscheleitkanälen gleichmäßige Linienbreiten. Die dokumentenechte, schnelltrocknende, säurefreie „Dry Safe" Tusche auf Wasserbasis ist nach dem Trocknen auf dem Papier nicht nur lichtecht, wisch- und wasserfest sondern auch optimal resistent gegenüber Chemikalien und Lösungsmitteln. Die Tusche eignet sich auch für Transparentpapier, da sie hier radierfähig ist. Der Stift kann bis zu acht Stunden offen liegen bleiben, ohne einzutrocknen. Um die Refill-Einheit auszuwechseln, wird einfach leicht auf die Aussparung am Stiftende gedrückt. Der Stift besitzt eine ergonomische Griffzone mit integriertem Rollstop und eine Sicherheitskappe mit einem hochwertigen Feingewinde, das versehentliches Öffnen verhindert.

This pen is an innovative writing and drawing pen that is very precise and easy to hold. Apart from its functional design and high-quality aluminium appearance, it is technically very sophisticated. The unique corrosion-resistant metal sleeve protects the point and keeps the pen steady when writing and drawing with rulers and templates. Together with specially adapted ink feed systems, the high-tech plastic point ensures that line width remains the same. Once it has dried on the paper, the water-based "Dry Safe" ink, which is permanent, dries fast and acid-free, is not only non-fading, non-smudging and waterproof, but also excellently resistant to chemicals and solvents. This ink is also suitable for transparent paper, as it can be erased here. The pen can be left open for up to eight hours without drying up. To exchange the refill unit simply apply light pressure to the recess at the end of the pen. The pen has an ergonomic grip zone with an integrated roll-stop and a safety cap with a high-quality fine-pitch thread, which prevents it being opened accidentally.

Jury ***Judges panel***

Ein bekanntes Produkt, mit edlen Materialien neu interpretiert, hervorragende Schreib- und Zeichenqualitäten.

A well known product, reinterpreted with high-quality materials, excellent writing and drawing qualities.

Hersteller ***Manufacturer***	**Vertrieb** ***Distributor***	**Design** ***Designer***
C. Josef Lamy GmbH Grenzhöfer Weg 32 D-69123 Heidelberg Tel. +49 62 21 84 31 04 www.lamy.de	C. Josef Lamy GmbH Grenzhöfer Weg 32 D-69123 Heidelberg Tel. +49 62 21 84 31 04 www.lamy.de	Franco Clivio Rietstraße 11 CH-8703 Erlenbach Tel. +41 19 12 06 60 Fax +41 19 12 06 62

Kugelschreiber „Pico" *ball point pen*

„Pico" ist ein neuartiger, technisch innovativer Pocket Pen, der sich klein machen kann. Im Taschenformat ist er gerade mal neun Zentimeter groß. Man kann ihn wie Feuerzeug, Schlüsselbund oder Lippenstift in der Tasche verschwinden lassen. Eine zum Patent angemeldete, technisch raffinierte Hubmechanik bewirkt, daß er sich vom kurzen Taschen- ins lange Schreibformat verwandelt. Ein sanfter Druck, „Pico" wächst auf zwölf Zentimeter und fährt zugleich die Schreibspitze aus. Noch ein leichter Druck, und er macht sich wieder klein. Innovativ ist auch die neuentwickelte Kompaktmine, die bis zu drei Kilometer lang schreibt. Die Schutzkappe der Mine hat gleichzeitig Adapterfunktion. Steckt man sie am hinteren Ende auf, so paßt die Mine auch in andere Lamy Kugelschreiber. „Pico" ist als Handschmeichler gestaltet. So ist er rundum gerundet, wodurch er angenehm in der Hand liegt und selbst ein empfindliches Taschenfutter nicht verletzt. Er ist in drei verschiedenen Oberflächen zu haben: in Pearlchrom, in glänzendem Chrom und in schwarzer Pulverbeschichtung. Das Logo ist im Hochrelief gestaltet, was ein Wegrollen des Stiftes auf glatten Flächen verhindert.

"Pico" is a new, technologically innovative pocket pen that can reduce its size. In its pocket size, it is just nine centimetres long. Liker a lighter, bunch of keys or lipstick, it can disappear into your pocket. A technically sophisticated lift mechanism, for which a patent is pending, helps it to change from its short pocket format to its long writing format. A light touch is all that is needed to make "Pico" grow to twelve centimetres and to make its ballpoint emerge. Another light touch, and it reduces in size again. The newly developed compact refill is also innovative. It will write for up to three kilometres. The protective cap on the refill also acts as an adapter. If it is clipped onto the other end of the refill, then the refill will fit into other Lamy ballpoint pens too. "Pico" fits comfortably into the hand. Every part of it is rounded, making it pleasant to hold and preventing it from damaging even the most sensitive pocket lining. It is available in three finishes: pearl chrome, polished chrome or with a black powder coating. The logo is embossed, thus preventing the pen from rolling on smooth surfaces.

Jury ***Judges panel***

Ein kleiner Zauberling, der sein wahres Wesen auf den ersten Knopfdruck entfaltet und auch noch funktioniert. Eine überraschende und in der Handhabung praktische, für die Tasche handliche Lösung. Handschmeichlerqualität beim Schreiben, das ist wichtig. Und eine verblüffende Verpackungslösung.

A magical little object that reveals its true character at the first touch of a button. And it works, too. A surprising solution, and one that is practical to use and handy for the pocket. When writing, it is important that a pen sits well in the hand. The packaging solution is stunning.

Hersteller ***Manufacturer***	**Vertrieb** ***Distributor***	**Design** ***Designer***
Do! Falkoner Allé 44, 4th DK-2000 Frederiksberg Tel. +45 40 19 23 35 designs@do-co.dk www.do-co.dk	Czechau Breite Straße 68 D-32657 Lemgo Tel. +49 52 61 39 39 Fax +49 52 61 1 55 15	Jeff Shore Falkoner Allé 44, 4th DK-2000 Frederiksberg Tel. +45 40 19 23 35 jpshore@yahoo.com

Kabelführung „Wire Snake" *cable organizer*

Wer hat sich an seinem Arbeitstisch nicht schon über Kabelsalat geärgert? „Wire Snake" ist eine ausgesprochen einfache, zum Patent angemeldete Lösung, um für Ordnung unter Kabeln zu sorgen. Die Idee für den Entwurf kommt von Kletterpflanzen, die im Dickicht andere Pflanzen umranken, oder von einer Schlange, die sich um etwas herumschlängelt. Ebenso schlängelt sich „Wire Snake" um Kabel von Computer, Telefon, Fax, Fernseher, Video, DVD-Player, Decoder und vielen anderen Geräten zu Hause und im Büro herum und räumt so auf. In der Handhabung ist die Kabelführung sehr simpel und benutzerfreundlich. Dank des flexiblen Materials wird sie einfach um die Kabel gebogen, so daß diese eng zu einem Bündel zusammen gehalten werden. Sind die Kabel zu lang, bildet man schlicht Schlaufen, die wieder von einer Schlange festgehalten werden. Sie läßt sich schnell wieder lösen und neu zusammenfügen. Darüberhinaus macht „Wire Snake" Spaß und fühlt sich angenehm in der Hand an. Die „Kabelschlange" aus elastischem Kunststoff ist in acht verschiedenen Farben erhältlich. Immer zwei sind in einer ungewöhnlichen transparenten Verpackung, die wie beschlagen wirkt, zu haben.

There are few of us who have not been annoyed by the tangle of wires under our desks. "Wire Snake", for which a patent is pending, is an outstandingly simple way of keeping wires together and tidy. The idea for the design was inspired by vines that wrap themselves around other plants, or by a snake that curls itself around something. In the same way, "Wire Snake" curls itself around the computer, telephone, fax, television, video, DVD player, decoder and other cables lying around in the office and the home, and tidies them. The cable organizer is very simple and user-friendly, Thanks to its flexible material, it is simply wound around cables, which are then held together tightly in a bundle. If the cables are too long, they can simply be made into loops, which in their turn can be held together by a snake. They are quick to undo again and to re-use. Moreover, "Wire Snake" is fun to use and fits comfortably into the hand. The "Wire Snake" of elastic plastic material is available in eight colours. They can be bought in packs of two, in an unusual frosted pack.

Jury ***Judges panel***

Absolut einfach in der Handhabung, extrem praktisch und notwendig, und dann auch noch farbenfroh und fröhlich. Und hübsch verpackt.

Absolutely simple to use, extremely practical and necessary, and then colourful and cheerful as well. And attractively packed.

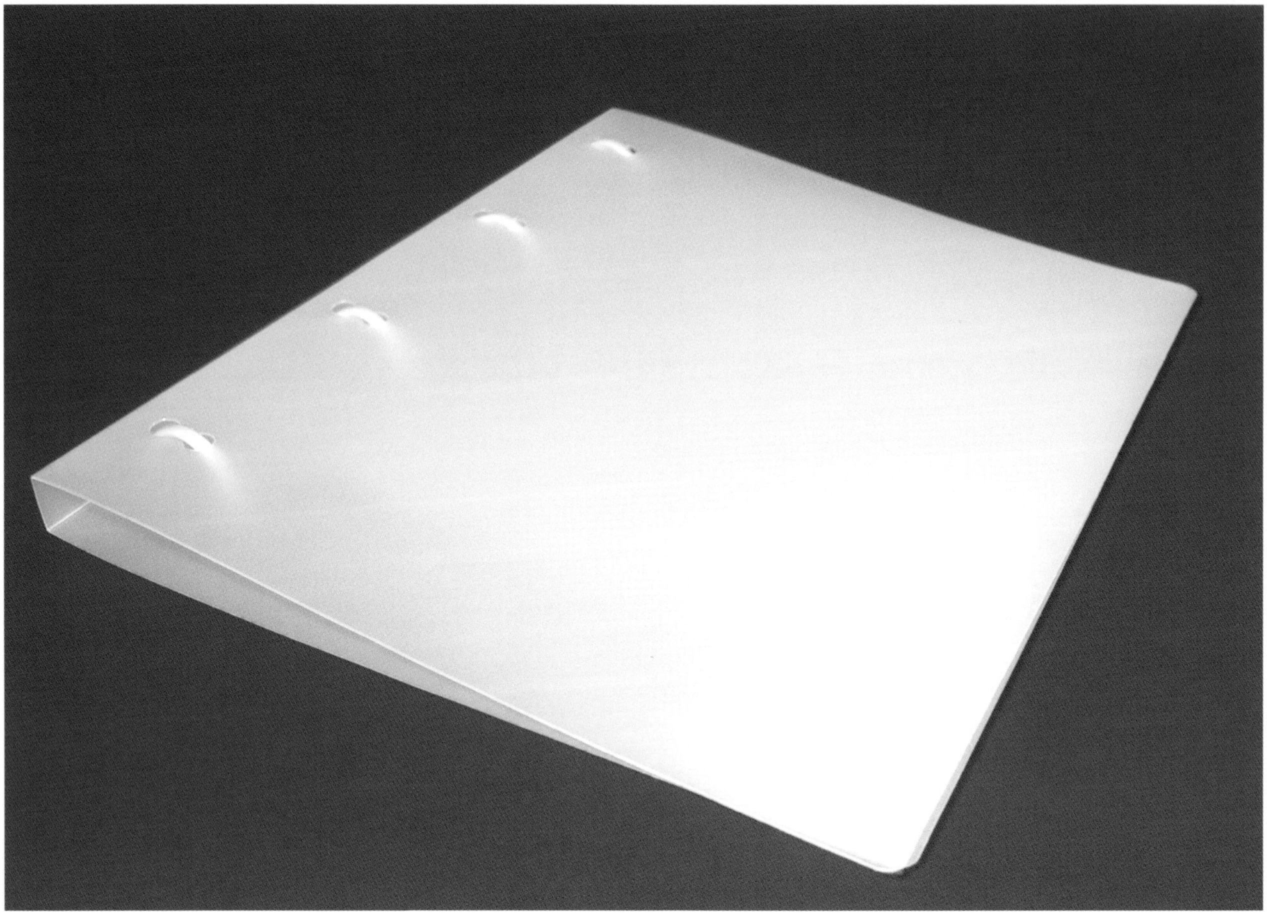

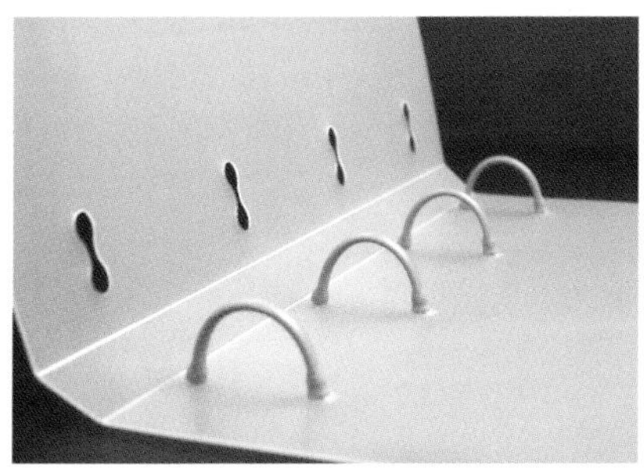

Hersteller ***Manufacturer***	**Vertrieb** ***Distributor***	**Design** ***Designer***
Lather Kommunikation Fronhäuser Straße 19 D-35102 Lohra-Damm Tel. +49 64 26 9 24 10 www.lather.de	Lather Kommunikation Fronhäuser Straße 19 D-35102 Lohra-Damm Tel. +49 64 26 9 24 10 www.lather.de	Modus Product Design GmbH Markgröninger Straße 61/1 D-70435 Stuttgart Tel. +49 7 11 8 70 21 05 Fax +49 7 11 8 70 21 06

Ringbuch „Polyring" *loose-leaf folder*

Polypin ist eine genial einfache, funktionale und komfortable Verbindung, die in den verschiedensten Polypropylen-Produkten eingesetzt werden kann. Sie funktioniert wie ein Druckknopf. Nun ist ein neues Mitglied zur Familie der wiederlösbaren Verbindungen hinzugekommen: der gebogene Polypin „Polyring". Er ersetzt die herkömmlichen Heftmappen, Schnellhefter und Spiralbindungen. Wie die anderen Polypins ist er leicht zu öffnen und immer wieder zu verschließen. Mappen mit „Polyring" bieten viele Vorteile. So lassen sich die Seiten umblättern, ohne daß diese zurückfallen. Es ist möglich, den Rücken zu beschriften. Damit können die Mappen übersichtlich in Regalen und Schränken aufbewahrt werden. Die Abstände der Lochung und die Anzahl der Pins lassen sich frei wählen. Nach Bedarf können unterschiedliche Blätter eingehängt werden, wie etwa Register oder ein Blatt mit Fächern, in denen Visitenkarten, Notizzettel und sogar Taschenrechner aufbewahrt werden können. Im Rücken läßt sich ein Schreibgerät unterbringen. Die Mappen können äußerst platzsparend gelagert werden, weil sie flach liegen und die Pins erst vor Gebrauch eingesetzt werden.

Polypin is a brilliantly simple, functional and comfortable binder that can be used in the widest variety of polypropylene products. It works like a press stud. A new member has now been added to the family of temporary binders: "Polyring", a curved polypin. It takes the place of conventional loose-leaf folders, fastener binders and spiral binders. Like the other Polypins, it is easy to open and can be closed again as often as necessary. There are many advantages to folders with "Polyrings". Pages can be turned, for example, without them falling back. The spines can be written on. This allows folders to be put away neatly in shelves and cupboards. Any distances between holes and number of pins can be chosen. If necessary, different sheets can be held together, such as indexes or a sheet with compartments for business cards, pieces of paper or even pocket calculators. There is room for a pen in the spine. The folders are extremely space-saving because they are laid flat, and the pins are not inserted until they are used.

Jury ***Judges panel***

Präsentationsmappe, schlicht, transparent und überzeugend, reduziert auf das Minimum, mehr braucht es nicht.

Presentation folder, plain, transparent and convincing, reduced to the minimum, there is no need for any more.

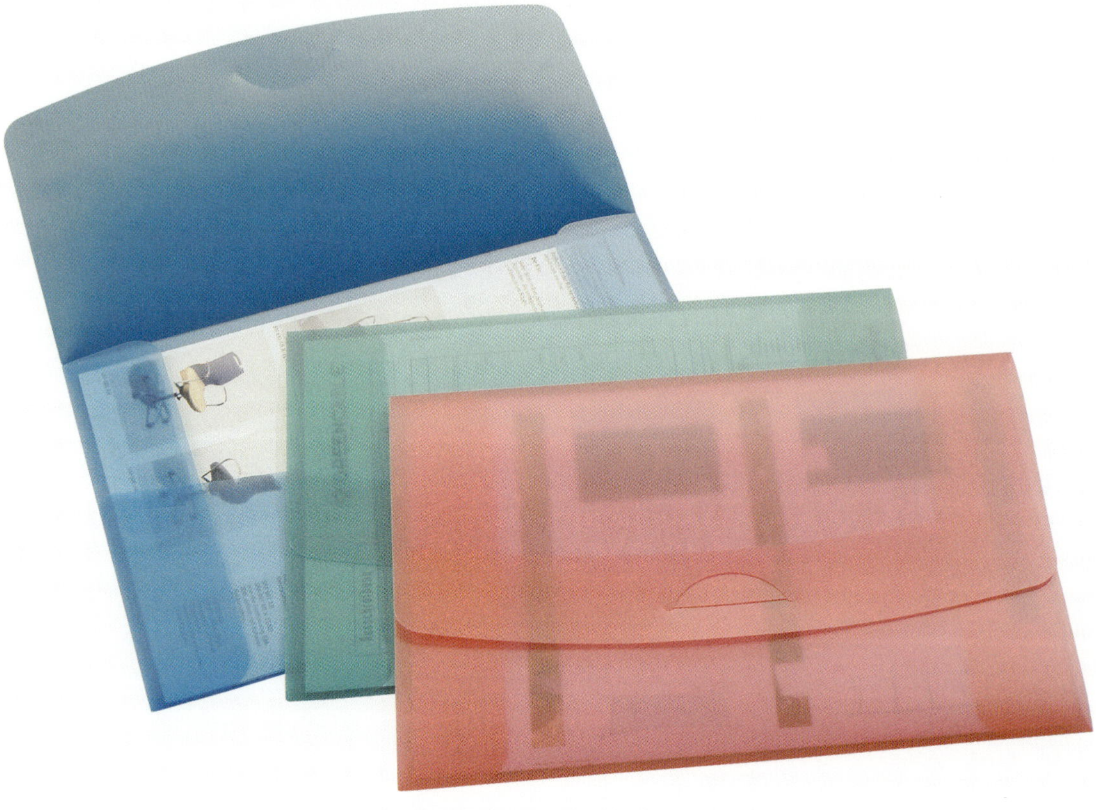

Hersteller ***Manufacturer***

Dataplus GmbH, Kratzert & Schrem
Hesselbühl 5
D-88630 Pfullendorf
Tel. +49 75 52 2 60 90
www.dataplus.de

Vertrieb ***Distributor***

Dataplus GmbH, Kratzert & Schrem
Hesselbühl 5
D-88630 Pfullendorf
Tel. +49 75 52 2 60 90
www.dataplus.de

Design ***Designer***

Heike Schnabel, Markus Schneider
Birkenstraße 14
D-73557 Mutlangen
Tel. +49 71 71 77 93 88
Fax +49 71 71 77 93 90

Sammelmappe 0–5 mm *0–5 mm folder*

Mit der Sammelmappe kann loses Schriftgut aufbewahrt und präsentiert werden. Dank ihrer Transparenz gibt sie leichten Einblick und läßt den Inhalt schnell erkennen. Die Mappe ist mit einem Steckverschluß ausgestattet, der einen schnellen Zugriff zum Inhalt erlaubt. Sie ist ganz aus einem Stück Polyboard, das heißt aus hundert Prozent Polypropylen, gefertigt, das den Anspruch erfüllt, unkompliziert recycelt und wieder verwendet werden zu können. Außerdem wird der anfallende Verschnitt beim Verarbeiten der Platten gesammelt und zu Granulat verarbeitet. Es gibt die Mappe in den Ausführungen von Null bis fünf Millimeter. Man kann aus den transparenten Farben natur, blau, rot, grün, rauch und opakem Weiß auswählen.

This folder allows loose pieces of paper to be filed and presented. Its transparency means that it is easy to see what is inside and to check its contents. The folder is equipped with a clip fastener which makes it easy to get at the contents. It is made entirely of one piece of Polyboard, i.e. one hundred per cent polypropylene, thus satisfying the criterion of being easy to recycle and capable of being reused. Moreover, any off cuts during the processing of the boards are collected and processed to make granules. The folder is available in models between zero and five millimetres. It is available in an untreated transparent version, in transparent blue, red, green and smoke, and in opaque white.

Jury ***Judges panel***

Im Endeffekt ist es ein bekanntes Produkt, das neu interpretiert worden ist. Was hier fasziniert, ist bei aller Kenntnis der Dinge, eine Lösung zu finden, die immer noch ein bißchen lapidarer daherkommt. Lapidar im positiven Sinne.

When all is said and done, a known product that has been reinterpreted. The fascinating thing here is to find a solution which, despite this prior knowledge, is a bit more succinct. Succinct in its positive sense.

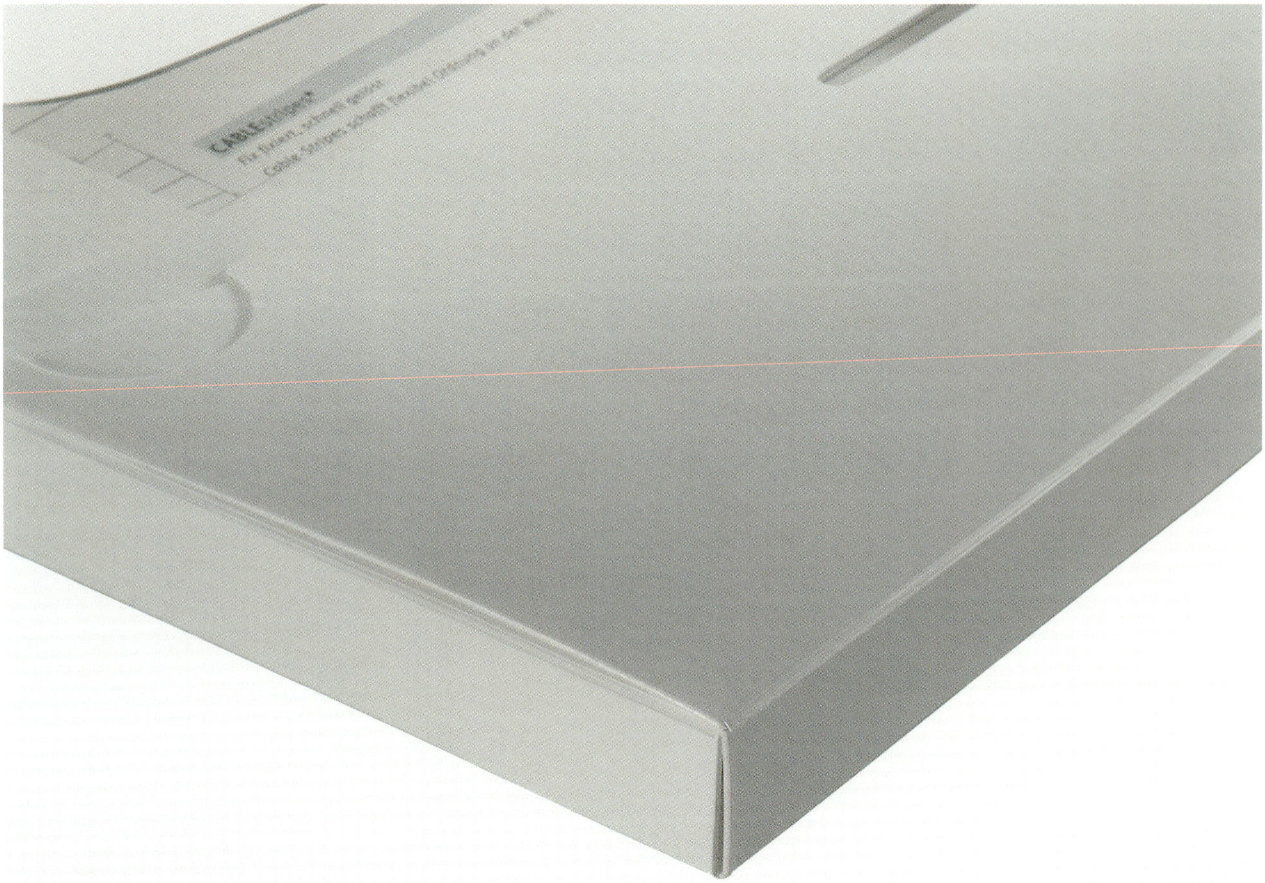
CABLEstripes®

Hersteller ***Manufacturer***	**Vertrieb** ***Distributor***	**Design** ***Designer***
Dataplus GmbH, Kratzert & Schrem Hesselbühl 5 D-88630 Pfullendorf Tel. +49 75 52 2 60 90 www.dataplus.de	Dataplus GmbH, Kratzert & Schrem Hesselbühl 5 D-88630 Pfullendorf Tel. +49 75 52 2 60 90 www.dataplus.de	Heike Schnabel, Markus Schneider Birkenstraße 14 D-73557 Mutlangen Tel. +49 71 71 77 93 88 Fax +49 71 71 77 93 90

Sammelmappe 5–20 mm *5–20 mm folder*

Die Sammelmappe nimmt alles an Unterlagen in Din A4-Format auf, was sonst nur schwer zu ordnen ist. Wie die schmaleren Varianten ist auch sie, um die Ressourcen zu schonen und die Umwelt zu entlasten, aus einem Stück leicht recycelbarem Polyboard hergestellt. Sie ist aber nicht nur umweltverträglich sondern auch besonders funktional und schön gestaltet. Die Mappen für eine bessere und attraktivere Ordnung im Büro aber auch für zuhause sind in den Farben natur-transparent und in undurchsichtigem Weiß erhältlich. Es gibt sie in den drei Füllhöhen fünf, zehn und zwanzig Millimeter. Die Ausführung mit zwanzig Millimeter ist dank 0,8 Millimeter Polyboard so stabil wie ein kleiner Aktenkoffer und eignet sich deshalb besonders für den Einsatz unterwegs.

This folder is the place to collect all those A4 documents that are otherwise difficult to find a home for. Like its slimmer versions, it is made of one piece of easily recyclable Polyboard, and so goes easy on resources and helps reduce pollution. However, its design is not just environmentally friendly; it is highly functional too, and attractively designed. These folders for better and more attractive order in the office or at home are available in an untreated transparent version or in opaque white. They are available for piles of documents five, ten and twenty millimetres thick. With its 0.8 millimetre thick Polyboard, the twenty millimetre version is as robust as a small briefcase, and hence is especially suitable for out-of-office use.

Jury ***Judges panel***

Langlebig, transparent, ordentlich, der einfache Stanzzuschnitt, aus einem Material, ohne Verbinder.

Long-lasting, transparent, tidy, simple punched shape, using just one material, without connecting materials.

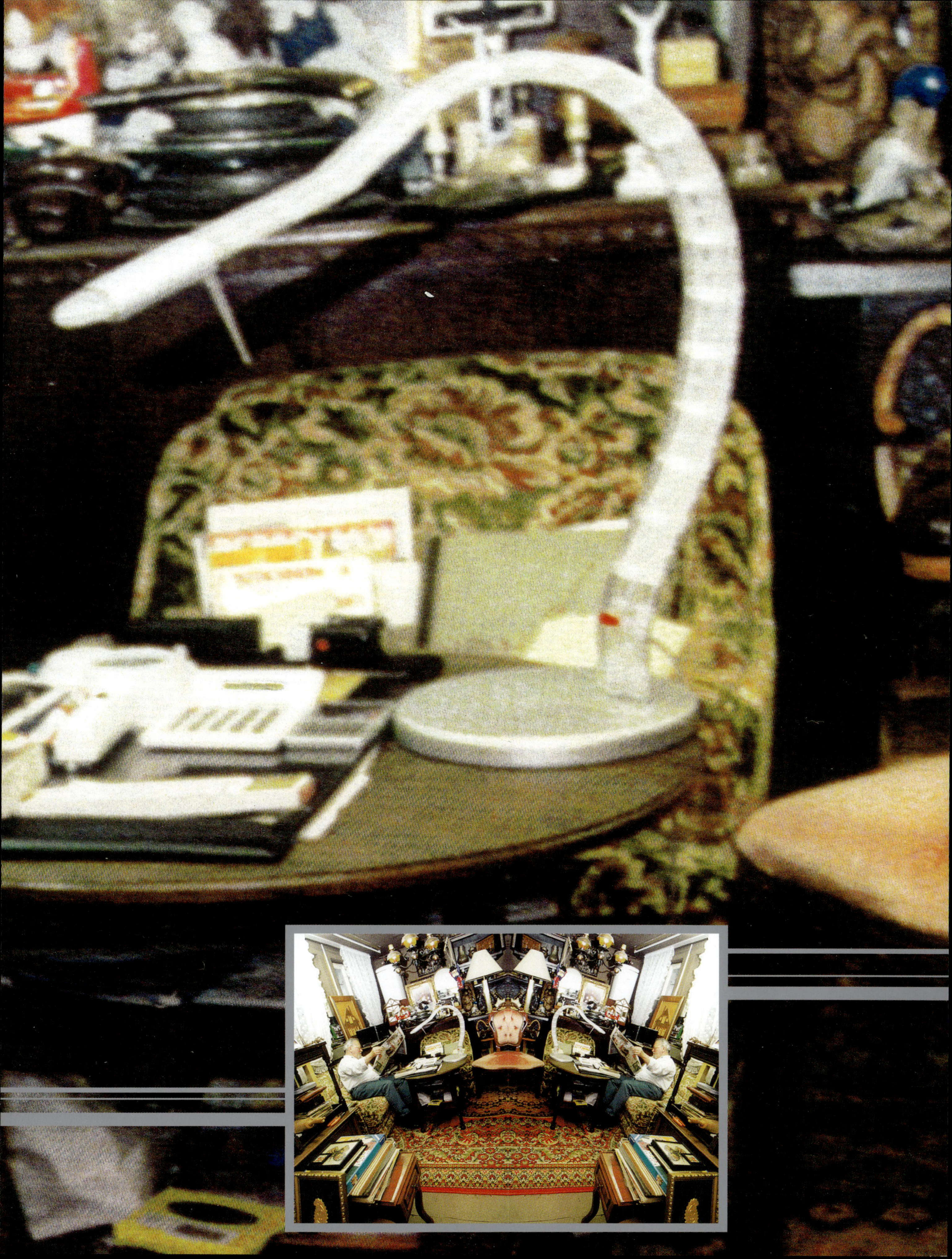

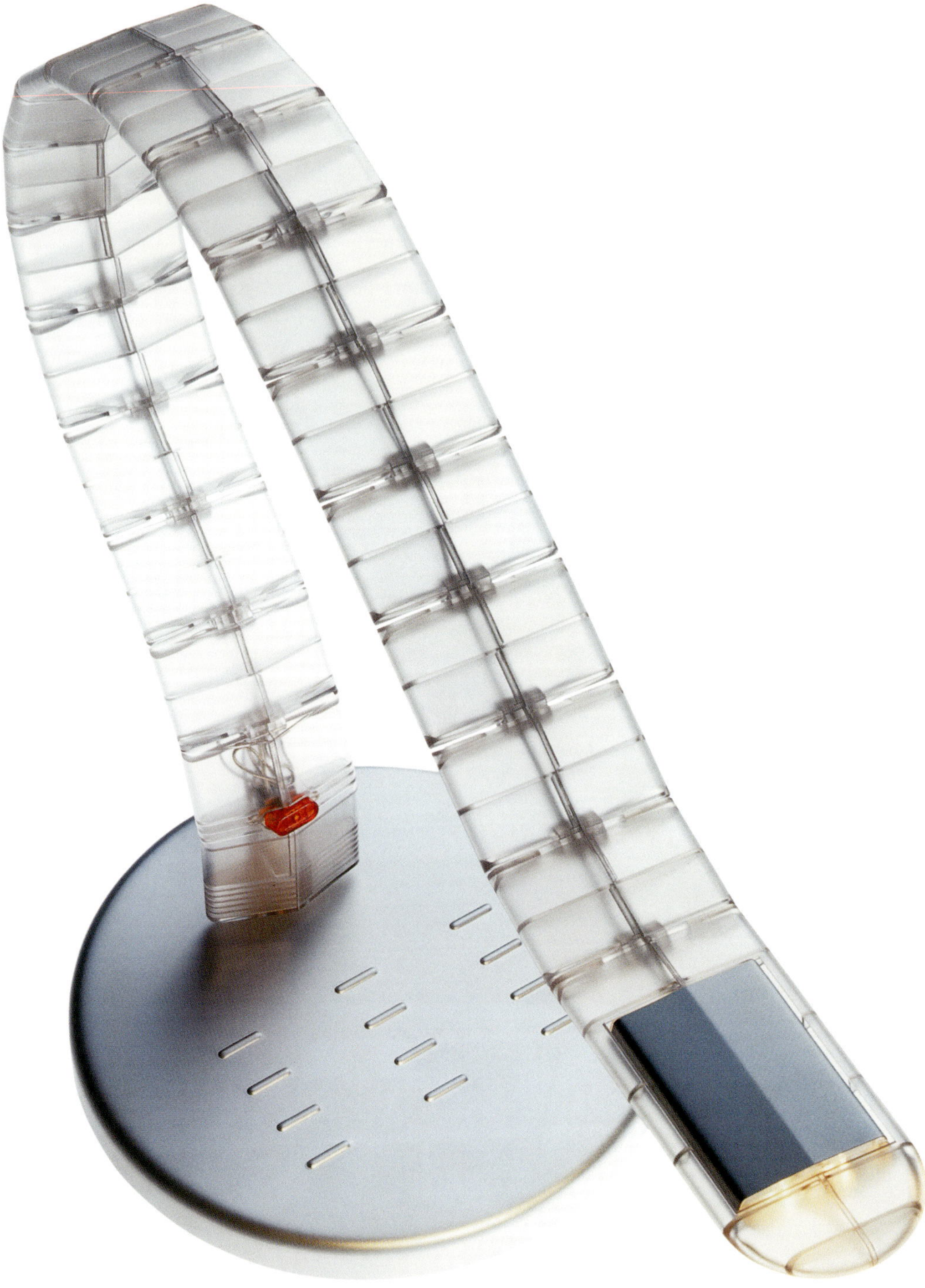

Hersteller ***Manufacturer***

Tobias Grau GmbH
Siemensstraße 35B
D-25462 Rellingen
Tel. +49 41 01 37 00
www.tobias-grau.com

Vertrieb ***Distributor***

Tobias Grau GmbH
Siemensstraße 35B
D-25462 Rellingen
Tel. +49 41 01 37 00
www.tobias-grau.com

Design ***Designer***

Tobias Grau
Werksdesign

Tischleuchte „Soon" *table lamp*

„Soon" besteht aus vielen gleichen Gliedern, wodurch sich die wie eine sich aufbäumende Schlange oder Wirbelsäule wirkende Tischleuchte ganz nach Belieben verstellen läßt. So erhält jede Leuchte ihre eigene charakteristische Form, die sich immer wieder verändern läßt. Die einzelnen Glieder wurden in transluzentem Kunststoff gefertigt, der den Blick auf das Innenleben freigibt. „Soon" ist mit einem Standfuß, mit einer Tischklemme oder einer Wandhalterung verwendbar. Sie ist in den Farben transparent und orange zu haben.

"Soon" is made up of many identical members, which means that the table lamp, which looks like a rearing snake or a spinal column, can be adjusted any way that is desired. Each lamp can thus take on a form of its own, which can be changed as often as is wanted. The individual members have been made of translucent plastic, thus making the inner workings of the lamp visible. "Soon" can be used with a stand, a desk clamp or a wall bracket. It is available in transparent and orange.

Jury ***Judges panel***

Bei Tobias Grau besticht der Mut zu wirklich innovativen Lösungen mit hohem technischen Aufwand. Bestechend ist neben der Lichtqualität die Konzentration auf kleine Lichtflächen. Faszinierend, wie er mit Kunststoff und mit Steckprinzipien nur über höchste Präzision von Werkzeugen zu sehr minimalen Lösungen kommt. Wenige in Europa schaffen das.

Tobias Grau's outstanding trait is his courage to go for truly innovative solutions that are technically highly intricate. What is striking apart from the quality of the light is its focus on small areas. It is fascinating how he arrives at extremely minimal solutions using plastic and a plug-in concept, simply on the basis of extremely precise tools. Only few European designers can do that.

Hersteller ***Manufacturer***	**Vertrieb** ***Distributor***	**Design** ***Designer***
Tobias Grau GmbH Siemensstraße 35B D-25462 Rellingen Tel. +49 41 01 37 00 www.tobias-grau.com	Tobias Grau GmbH Siemensstraße 35B D-25462 Rellingen Tel. +49 41 01 37 00 www.tobias-grau.com	Tobias Grau Werksdesign

Hängeleuchte „Project X" *pendant lamp*

„Project X" ist eine höhenverstellbare Hängeleuchte mit doppelter Lichtwirkung. Vollkommen blendfreies Licht wird durch opales Glas in großer Menge nach unten abgegeben und läßt gleichzeitig farbige Gläser nach oben leuchten.
Die Gestaltung der Leuchte zeichnet sich durch eine elegante flache Scheibe aus, eine pure schlichte Form. Sie ist in den Farben schwarz, purpur, orange und transparent zu haben.

"Project X" is a height-adjustable pendant lamp with a dual lighting effect. Large amounts of completely glare-free light are shone downwards through opal glass, while at the same time the light causes coloured glass to shine upwards. The lamp is designed as an elegant flat disk, a pure, modest form. It is available in black, purple, orange and transparent.

Jury ***Judges panel***

Das Ufo für das Jahr 2020. Einfache Formgebung, saubere Verarbeitung, bestechende Materialverarbeitung, hochwertig, schöne Lichtqualität. Es leuchtet auch nach oben, indirekt; gibt oben und unten Helligkeit. Schöne Diskusform.

The UFO for 2020. Simple form, perfect finishing, attractive use of materials, high quality, emits a beautiful light. It also shines light upwards, indirectly; illuminates below and above. Nice discus form.

Hersteller ***Manufacturer***	**Vertrieb** ***Distributor***	**Design** ***Designer***
Artemide S.p.A. Via Bergamo 18 I-20010 Pregnana Milanese Tel. +39 02 93 51 81 www.artemide.com	Artemide GmbH Hans-Böckler-Straße 2 D-58730 Fröndenberg Tel. +49 23 73 97 50 www.artemide.com	Natali Crasset c/o Artemide S.p.A. Via Bergamo 18 I-20010 Pregnana Milanese

Hängeleuchte „Ierace" *pendant lamp*

„Ierace" ist eine nicht allzu große, neue Leuchtstofflampe, die sich durch hohe Leistung und warme Lichtfarbe auszeichnet. Die technologische Forschung im Bereich der Lichtquellen wurde hier ganz in den Dienst des Sehkomforts und der Ästhetik gestellt. So hat sie direkte und indirekte Lichtverteilung. Der optisch leichte Eindruck des Designs – „Ierace" scheint zu schweben – wird durch die Kabel, die zur Deckenaufhängung führen und gleichzeitig als Netzkabel dienen, noch unterstrichen. Der Rahmen ist aus lackiertem Aluminium gefertigt. Der äußere Lichtverteiler besteht aus transparentem, mundgeblasenem Glas, die zwei inneren Lichtverteiler sind wie die Rosette aus Polykarbonat hergestellt.

"Ierace" is a not over-sized, new fluorescent lamp, characterized by high performance and a warm light. Here, technological research into light sources has been devoted entirely to making eyes feel comfortable and creating an aesthetic effect. It distributes light directly and indirectly. The visually light appearance of the design – "Ierace" seems to float – is emphasized even more by the cables, which lead to the ceiling suspension unit and serve at the same time as power cables. The frame is made of painted aluminium. The outer light distributor is made of transparent, blown glass, while the two inner light distributors, like the rosette, are made of polycarbonate.

Jury ***Judges panel***

Eine Deckenleuchte mit archetypischen, puristischen Formen und Einfachstmaterialien kombiniert. Es bestricht, wie hier mit Licht, Glas, Metall und der Abhängung durch feine Drähte umgegangen wird.

A ceiling light combined with archetypical, purist forms and the simplest of materials. It is captivating how this design makes use of light, glass, metal and fine wires for suspension.

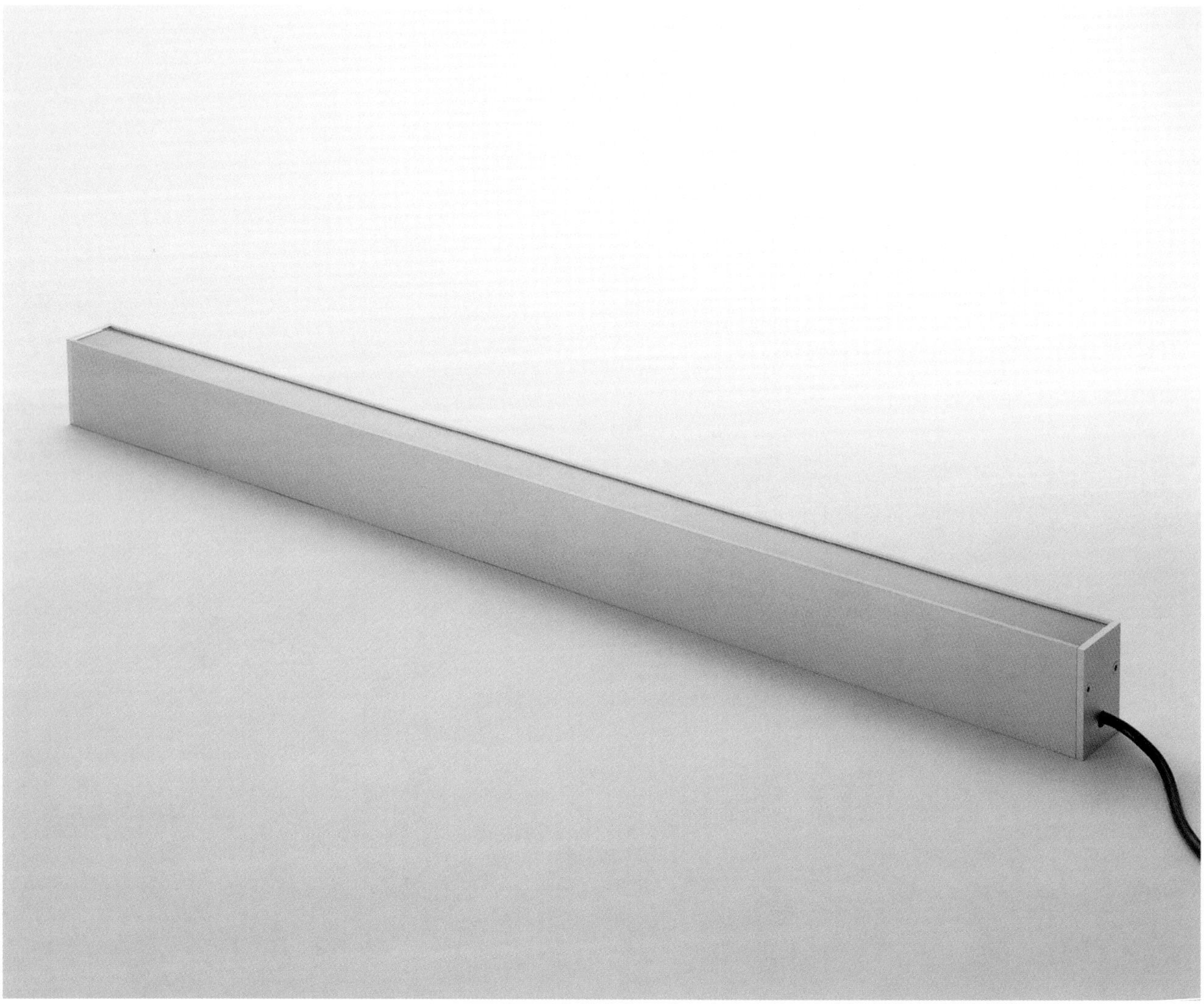

Hersteller ***Manufacturer***	**Vertrieb** ***Distributor***	**Design** ***Designer***
LRG. Licht + Raum AG Hinterer Schermen 44 CH-3063 Ittingen Tel. +41 31 9 24 77 88 LRG@bluewin.ch	Astra Licht- und Deckensysteme Vertriebs GmbH Talstraße 19 D-70825 Korntal-Münchingen Tel. +49 7 11 8 39 50 00	Werksdesign

Tischleuchte „U70" *table lamp*

„U 70" – das ist Licht als schlichte Linie, als Lichtstab umgesetzt. Mit einem Querschnitt von gerade mal sieben mal viereinhalb Zentimeter ist die Erscheinung der Leuchte lang und schmal. Dabei gibt es sechs unterschiedliche Längen von 60 Zentimeter bis drei Meter. Das Design ist gekennzeichnet durch strenge formale Reduktion. Damit hält sich die Leuchte gegenüber der Architektur bescheiden im Hintergrund und unterstützt dafür deren räumliche Qualitäten. Die Leuchte ist ausgesprochen vielseitig einsetzbar. Die gewünschte Wirkung, die erstrebte Stimmung bestimmt Montage und Anwendung. So kann sie als Pendelleuchte von der Decke hängen, direkt an die Wand oder mit Distanzwinkel sowie unter Putz montiert werden. Da sie mit Kabel und Stecker ausgestattet ist, läßt sie sich auch mobil anwenden, beispielsweise an die Wand stellen, auf Tisch oder Boden legen. Die Leuchte besteht aus stranggepreßtem, farblos eloxiertem Aluminiumprofil. Außerdem hat sie eine UV-hemmende Abdeckung aus opaler Polycarbonatfolie.

"U 70" – a synonym for light as a plain line, put into practice as a neon tube. With a cross-section of just seven by four-and-a-half centimetres the lamp appears to be long and slim. It is available in six different lengths, from 60 centimetres to three metres. The design stands out by virtue of its strict formal reduction. This means that the light keeps itself modestly in the background, supporting the spatial qualities of its architectural setting.The light can be used in a wide variety of ways. The effect or mood desired will determine how it is mounted or used. It can hang from the ceiling as a pendant lamp, for example, or be mounted directly on the wall, either with a spacer bracket or flush with the wall surface. As it is equipped with cable and a plug, it can also be picked up and moved around: leaned against the wall, for example, or laid on a table or the floor. The lamp is made of extruded, colourlessly anodized aluminium section. It also has a UV-retarding cover of opal-coloured polycarbonate sheeting.

Jury ***Judges panel***

Ganz simples Produkt, bei dem der Benutzer das Gefühl hat, daß er die Verwendungsmöglichkeiten selbst finden muß. Die Lampe kann gelehnt, gelegt, gestellt werden. Die Möglichkeiten sind kaum vordefiniert. Sehr hohe Mobilität.

Very simple product, giving the user the feeling he has to find out how to use it himself. The lamp can be leaned, laid, positioned. The possibilities are very vaguely defined in advance. Extremely high mobility.

Hersteller ***Manufacturer***	**Vertrieb** ***Distributor***	**Design** ***Designer***
Sammet Moebelkonzepte Ellenbergstraße 19 D-71729 Erdmannhausen Tel. +49 71 48 16 24 77 frank.sammet@bueroform.net	Sammet Moebelkonzepte Ellenbergstraße 19 D-71729 Erdmannhausen Tel. +49 71 48 16 24 77 frank.sammet@bueroform.net	Frank Sammet Sonnenbergstraße 2 D-71723 Großbottwar Tel. +49 1 71 1 95 33 97

Systemring „S8" *system ring*

Der Systemring hat die Funktion, Arbeitsflächen auf Tischen, zumeist auf Schreibtischen, besser zu organisieren. Der aus Aluminium gefertigte Ring wird in einem 68 Millimeter großen Loch von der Unterseite mit einem Gewindering verspannt. Hierfür sind Tischplatten mit einer Stärke von 18 bis 32 Millimeter geeignet. Die weichen Formen, das reine hochwertige Material sowie der hohe Nutzen charakterisieren das Produkt. Der Systemring ist vielseitig einsetzbar, es können damit verschiedenste Gebrauchsgegenstände auf dem Tisch organisiert werden. Die einfachste Nutzung ist als Kabeldurchgangsbuchse. Mit verschiedenen Einsätzen läßt sich der Systemring auch noch anders verwenden. So kann er beispielsweise Leuchten aufnehmen, etwa die Leuchte Tolomeo von Artemide. Dabei wird das Stromkabel gleichzeitig mit durch den Tisch geführt. Es können aber auch Telefon- oder Monitorarme für Flachbildschirme eingesetzt werden. Schließlich lassen sich auch Reagenzgläser einstecken, die als kleine Blumenvasen dienen, oder aber der Einsatz wird als Kerzenhalter verwendet.

The purpose of this system ring is to organize working space on tables (usually on desks) better. Made of aluminium, the ring is placed in a 68 millimetre hole in the desk top, and secured from below with a threaded ring. It is suitable for desk tops between 18 and 32 millimetres thick. The product is characterized by soft forms, pure, high-quality material and high utility value. The system ring can be used in a variety of ways, allowing many different desk-top utensils to be organized. Its simplest use is as a hole for cables to pass through. Various inserts allow the system ring to be used in other ways, too. It can act as a mount for lamps, for example, such as the Tolomeo lamp by Artemide. In this case, the power cable is also passed through the desk at the same time. Telephone arms or monitor arms for flat-screen monitors can also be mounted. Finally, test tubes can be inserted and serve as small vases, or the hole can be used as a candle holder.

Jury ***Judges panel***

Ein klassischer Kabeldurchlaß für Schreib- oder Bürotische mit Zusatzfunktion versehen durch Steckmodule, die einerseits Lampen aufnehmen, weiterhin aber Kabelführung haben oder wenn keine Kabelführung notwendig ist, können sie eine versteckte Blumenvase aufnehmen und bilden damit auch eine Art Dekoration.

A classic cable duct for desks or office tables, given additional functions through modular inserts, which can either act as a mount for lamps but still allow cable to pass through or, if a cable duct is not necessary, can hold a concealed vase and thus perform a kind of decorative function.

FISKARS
FISKARS

Hersteller ***Manufacturer***	**Vertrieb** ***Distributor***	**Design** ***Designer***
Fiskars Consumer Oy Ab FIN-10330 Billnäs Tel. +3 58 19 27 77 21 Fax +3 58 19 23 09 86 www.fiskars.fin	Fiskars Deutschland GmbH Weststraße 42 D-40721 Hilden Tel. +49 21 03 5 89 00 Fax +49 21 03 58 90 45	Olavi Lindén, Markus Paloheimo Werksdesign

Gelenkgartenscheren „Power-Lever® Bypass und Amboß"
articulated garden shears

„Bypass und Amboß" zeichnen sich durch einen ungewöhnlichen Gelenkmechanismus aus. Die parallele Griffbewegung erlaubt, daß die beiden stärksten Finger auch mehr Arbeit verrichten. Der einzigartige patentierte Gelenkmechanismus macht nicht nur die Schere leicht und kompakt, sondern erhöht auch die Schneideleistung um 35 Prozent. Die Finger- und Armmuskulatur wird durch ein durchdachtes ergonomisches Design geschont. Die Scheren liegen gut in der Hand, passen sowohl in kleinere wie auch mittlere Hände. Sie wurden so gestaltet, daß Links- wie Rechtshänder sie gleichermaßen benutzen können. Die Oberfläche hat eine angenehme Haptik. Der Verschlußmechanismus funktioniert mit einer einfachen Berührung und schützt so die Klingen und verhindert Unfälle.

"Bypass and Amboss" are characterized by an unusual articulated mechanism. The parallel grip movement means that the two strongest fingers do more of the work. The unique, patented articulated mechanism not only makes the shears light and compact, but also increases their cutting power by 35 per cent. The well thought-out ergonomic design saves work for the finger and arm muscles. The shears fit well into the hand, and are suitable both for small and medium-sized hands. They have been designed so that they can be used equally well by left and right-handers. Their surface is pleasant to the touch. A simple touch is all that is needed to engage the locking mechanism, which protects the blades and prevents accidents.

Jury ***Judges panel***

Eine optisch einwandfreie und klare Linienführung, haptisch angenehm zu greifen, schmeichelt der Hand. In der Funktion dann eine überraschende Stärke, die durch den zusätzlichen Hebeleffekt entwickelt wird. Überzeugendes Gartengerät für große und kleine Hände. Funktionelle Verpackung.

Optically perfect, clear lines, pleasant to the touch, fits nicely into the hand. Surprisingly strong when it is used, its strength coming from the additional leverage effect. Convincing garden tool for large and small hands. Functional packaging.

Hersteller ***Manufacturer***	**Vertrieb** ***Distributor***	**Design** ***Designer***
Walz GmbH & Co Im Lehrer Feld 6 D-89081 Ulm Tel. +49 7 31 96 27 70 www.walz-ulm.de	Walz GmbH & Co Im Lehrer Feld 6 D-89081 Ulm Tel. +49 7 31 96 27 70 www.walz-ulm.de	Michael Leggewie Oberer Eschachweg 6 D-89284 Pfaffenhofen Tel. +49 73 02 31 97 design@vision-produktgestaltung.de

LED Taschenlampe „Techno" *LED pocket torch*

Die Taschenlampe zeichnet sich durch elegantes, sachliches und funktionales Design, innovative Technik und schöne Materialkombination aus. So hebt sich die Kappe aus Edelstahl sehr schön vom Griff aus Softkunststoff ab. Dieser ist genoppt und damit gut zu greifen. Durch eine microprozessorgesteuerte LED hat „Techno" mehrere Schaltfunktionen. So kann man die Taschenlampe dimmen, blinken, auf Dauerlicht oder Standby stellen sowie abschalten. Herkömmliche LED-Lampen benötigen mindestens drei Knopfzellen, was zusätzliche Kosten verursacht. „Techno" dagegen braucht nur eine einzige Batterie, die überall für wenig Geld gekauft werden kann.

The distinctive features of this torch are its elegant, sober and functional design, its innovative technology and beautiful combination of materials. The stainless steel cap contrasts very beautifully with the soft plastic grip. The latter is nippled, making it easy to grip. Thanks to a microprocessor-controlled LED, "Techno" has several switching functions. The torch can be dimmed, shine intermittently or permanently, be put on stand-by or switched off. Traditional LED torches need at least three round cells, causing additional costs. By contrast, "Techno" needs just one battery, which can be bought inexpensively anywhere.

Jury ***Judges panel***

Neue Leuchtstoff-Technologie, elegant, innovativ und mit verschiedenen Funktionen überzeugend umgesetzt. Die tausendste Lampe und trotzdem innovativ. Sehr schön gemacht. Es ist auch ein Handschmeichler. Normalerweise sind die Taschenlampen noch ein bißchen kleiner. Aber als Klassiker ist sie unschlagbar.

New lighting technology, elegant, innovative and combined with a series of sensible functions. The thousandth torch, but nevertheless innovative. Very beautifully made. It fits nicely in the hand, too. Normally, pocket torches are a little smaller. But as a classic, this one cannot be beaten.

Hersteller ***Manufacturer***	**Vertrieb** ***Distributor***	**Design** ***Designer***
Protrade Europe Pforzheimer Straße 63 D-75449 Wurmberg Tel. +49 70 44 94 51 20 www.protrade.de	Protrade Europe Pforzheimer Straße 63 D-75449 Wurmberg Tel. +49 70 44 94 51 20 www.protrade.de	Lutz Gathmann Nagelsweg 41 D-40474 Düsseldorf Tel. +49 2 11 45 29 19 www.lutz-gathmann.de

Teelicht „Charm" *night-light*

„Charm" ist ein vielseitiges Produkt für zuhause, für die anspruchsvolle Hoteleinrichtung oder den modernen Reisenden. Zum einen ist es ein schlichtes Teelicht, das auch als Notlicht dient, zum anderen kann der Grundbehälter vielfältig vom Taschenaschenbecher bis zur Pillendose genutzt werden. „Charm" sticht durch seine elegante, schlichte Form hervor. Der umlaufende Gummiring erleichtert die Handhabung. Der drehend gelagerte Deckel wird durch Magnete so dicht verschlossen, daß das Kerzenlicht einfach durch Schließen gelöscht werden kann. Zudem dient er im ausgedrehten, das heißt geöffneten Zustand als Griff. Das Material des Teelichts ist Zinkdruckguß.

"Charm" is a versatile product for the home, as part of up-market hotel furnishing, or for the modern traveller. On the one hand it is a plain night-light, which can also be used as an emergency light, and on the other hand its holder base can be used in many different ways, from pocket ashtray to pillbox. The distinctive feature of "Charm" is its elegant, unassuming form. The rubber ring running around it makes it easier to use. The lid, which is opened and closed by rotating it to the side, is sealed so tightly by magnets that the candle is extinguished simply by closing it. When turned out, i.e. open, it serves as a handle. The night-light is made of die-cast zinc.

Jury ***Judges panel***

Einfach, schön versteckte Mechanik, mit bestechender Funktion, reisefähig, schöne Verpackung für ein althergebrachtes Lichtutensil. Perfekte Qualität. Ein Hightech-Teil, mechanisch ausgedreht. Im Vergleich zu unzähligen, mehr oder weniger festlich wirkenden Teelichtvarianten eine bestechend simple Lösung, auf das Wichtigste reduziert.

Simple, well concealed mechanism, which functions outstandingly, good for journeys, attractive packaging for such a traditional light. Perfect quality. A high-tech item, opened mechanically. Compared with the so many night-light variants whose appearance is more or less formal, this solution is captivatingly simple, reduced to the bare essentials.

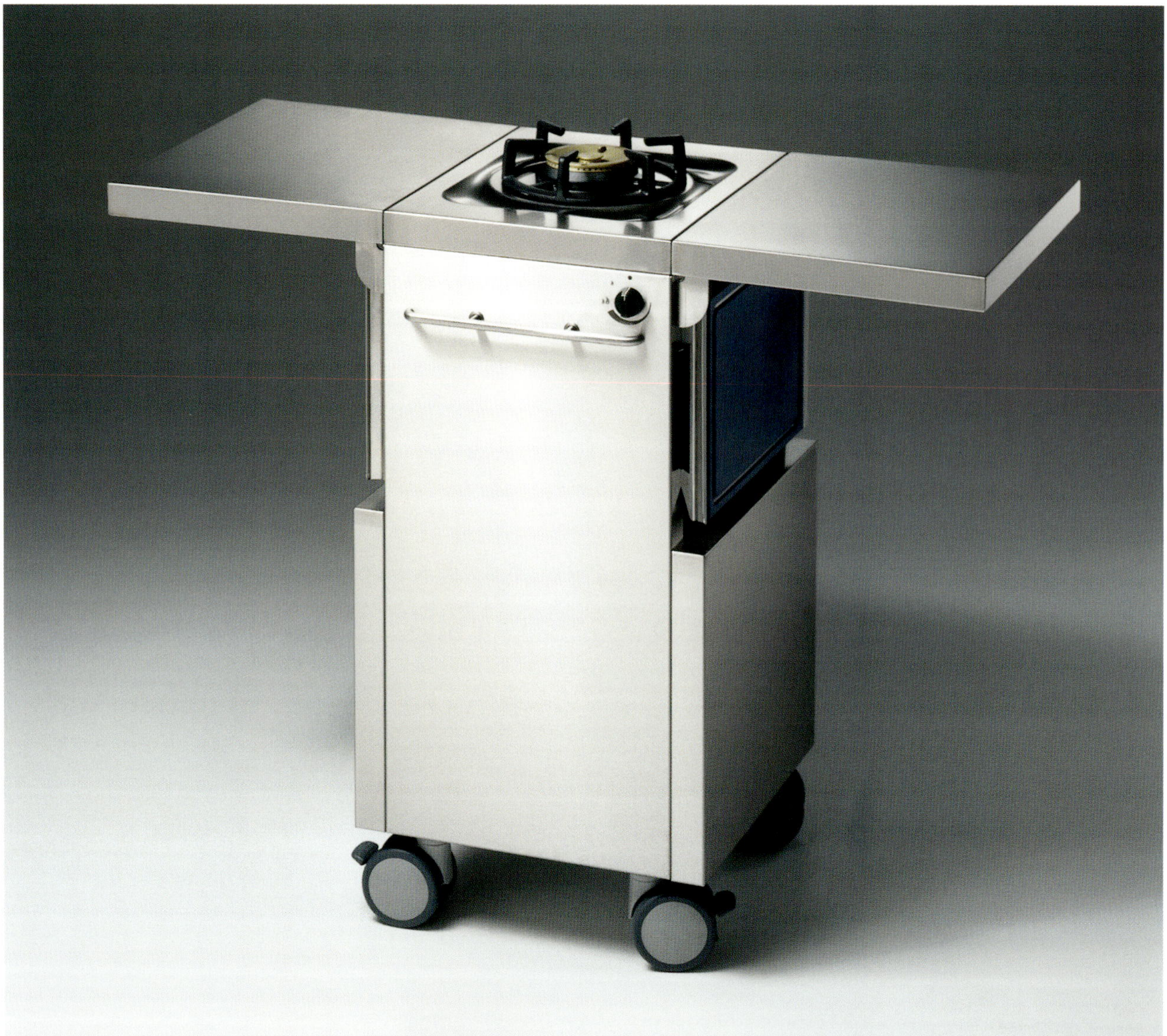

Hersteller ***Manufacturer***	**Vertrieb** ***Distributor***	**Design** ***Designer***	
Rieber GmbH & Co. KG Hoffmannstraße 44 D-72770 Reutlingen Tel. +49 71 21 51 86 40 www.rieber.de	Rieber GmbH & Co. KG Hoffmannstraße 44 D-72770 Reutlingen Tel. +49 71 21 51 80 www.rieber.de	Waltraud Claar Untere Bergstraße 3 D-71334 Waiblingen Tel. +49 71 51 3 45 54 Fax +49 71 51 50 86 26	Werksdesign

Mobiles Küchenmodul „Claarissimo®" *mobile kitchen module*

Der erste mobile Gasherd mit Profitechnik in kompakter Form kann bei Party, Event oder Empfang, drinnen wie draußen, im professionellen Gastronomie- und Hotelbetrieb genauso wie zu Hause eingesetzt werden. Dafür gibt es zwei unterschiedliche Größen: eine für den privaten und eine für den professionellen Gebrauch. Auf dem mobilen Gasherd kann sowohl mit Topf und Pfanne als auch mit Wok gekocht werden. Da er mit einer Gasflasche betrieben wird, ist er völlig autark, also ideal für den Garten. Seitlich bieten ausklappbare Arbeitsflächen ausreichend Platz zur Essensvorbereitung. Auf Wunsch gibt es dazu passende Schneidbretter in Kunststoff oder Holz. Doppellenkrollen sorgen für einen leisen und leichten Lauf, ausziehbare Griffe für einfachen Transport auch über unebene Flächen oder Stufen. Der Gasherd zeichnet sich durch seine auf das Wesentliche reduzierte, sachlich-elegante und funktionale Gestaltung aus. Der klar geformte Korpus besteht aus hochwertigem rostfreien Edelstahl. Neben dem Gasherd stehen noch vier weitere Module zur Auswahl: „Varithek®" zum Kochen, Grillen und Speisen warmhalten, Handwaschbecken, ein Gläserspüler und ein Schubladenmodul.

The first mobile gas cooker to combine professional technology with compactness can be used at parties, events or receptions, inside and out, in the professional catering and hotel trade just as well as at home. It is available in two different sizes: one for private and one for professional use. Cooking can be done on the mobile cooker both with pots and pans and with a wok. As it works with bottled gas, it is completely self-contained, and so is ideal for the garden. To the sides, work surfaces can be folded out, providing sufficient room to prepare food. If desired, there are matching plastic or wood cutting boards. Double guide rolls ensure that it runs quietly and easily, and extendable handles allow it to be transported easily even over uneven surfaces or steps. The distinctive features of the gas cooker's design are that it has been reduced to the bare essentials, and is sober, elegant and functional. The body with its clear lines is made of high-quality stainless steel. Four other modules are available apart from the gas cooker: "Varithek®" for cooking, grilling and keeping food warm, a hand washbasin, a glass washing machine and a plate dispenser module.

Jury ***Judges panel***

Dieser Einflammen-Gaskocher entpuppt sich, wenn er seine Flügel entfaltet, als eine außerordentlich charmante Kochhilfe für den Outdoor-Bereich, an der das Kochen mit verschiedenem Zubehör mehr Spaß macht. Ein geradezu perfektes mobiles Gartenkochgerät, technisch gut durchdacht, auch die Materialkombination von Edelstahl und Holz als Beibretter ist sehr überzeugend.

When it opens its wings, this single-flame gas cooker reveals itself to be an extraordinarily charming cook's aid for the outdoor area, which increases the fun of cooking with its various accessories. A practically perfect mobile garden cooking appliance, technically well thought-out, and the combination of stainless steel and wood as cutting boards is very convincing, too.

www.leggero.com

Leggero

Hersteller ***Manufacturer***	**Design** ***Designer***
Brüggli Produktion & Dienstleistung Hofstraße 5 CH-8590 Romanshorn Tel. +41 71 4 66 94 94 www.leggero.com	Chrétien + Apothéloz Lägernstraße 27 CH-8037 Zürich Tel. +41 1 3 61 51 47 www.industrialdesigners.ch

Fahrradanhänger „Leggero Cuatro" *bicycle trailer*

Mit „Leggero Cuatro" wurde ein Fahrzeugsystem – die Weiterentwicklung eines erfolgreichen Vorgängermodells – geschaffen, das Familien zuverlässig durch die ersten Jahre ihrer Kinder begleitet. In Sicherheit und Design setzt es neue Maßstäbe, nicht zuletzt durch seine Multifunktionalität. So kann es dank verschiedenem Zubehör nicht nur als Fahrradanhänger, sondern auch zum Wandern, Einkaufen, für Warentransporte, Spazierengehen, Joggen, Skaten, für den Urlaub und als Kinderwagen eingesetzt werden. Der Fahrradanhänger ist leicht zusammen zu klappen, das Zubehör einfach zu montieren. Er ist in gelb, rot sowie blau und in zwei Ausführungen erhältlich: Einsitzer und Zweisitzer für Kinder von ein bis sechs Jahren. Ein neues, einzigartiges ergonomisches Sitzsystem – ein variabel verstellbarer Alurohrrahmen mit abnehmbarem Dreischichtnetz und Dreipunktgurt – sorgt für Komfort und für sicheren Kindertransport. Hohe Verarbeitungsqualität und innovative hochwertige Werkstoffe bieten Langlebigkeit und einen sehr guten Kundennutzen. So besteht das Chassis aus ABS, verstärkt durch Seitenteile aus Aluminium, die Deichsel aus Chromstahlrohr.

"Leggero Cuatro" is a vehicle system – the further development of a successful predecessor model – that reliably accompanies families through their children's first years. In terms of safety and design it sets new standards, not least because of its multi-functionality. Thanks to various accessories it can be used not only as a bicycle trailer, but also for hiking, shopping, to transport goods, for walking, jogging, skating, for holidays and as a push-chair. The bicycle trailer is easily folded down, and the accessories are easy to attach. It is available in yellow, red and blue, and in two versions: with one or two seats for children between one and six years old. A new, unique ergonomic seating system – an adjustable frame of aluminium tubing with a removable three-ply net and a three-point seat belt – makes it comfortable and ensures that children are transported safely. High-quality finishing and innovative high-quality materials provide longevity and very good customer benefit. The chassis is made of ABS, reinforced by side elements of aluminium, and the drawbar is made of chromium-plated tubular steel.

Jury ***Judges panel***

Mit diesen Geräten werden die Kinder in einem mobilen Sinne ernst genommen, bekommen ihre eigene Umwelt und Intimität. Umweltfreundlich mobil auch bei ungünstigen Witterungsbedingungen. Reichhaltigkeit der Materialien, harte und weiche Materialien, schöne Details, optimale Beweglichkeit. Durch geringfügiges Umbauen eigenfahrbar.

These appliances show that children are taken seriously in a mobile sense, and are given their own environment and intimacy. They offer environment-friendly mobility even in inclement weather. A wealth of materials, hard and soft, nice details, excellent mobility. A few adjustments, and it can be driven independently of a bicycle.

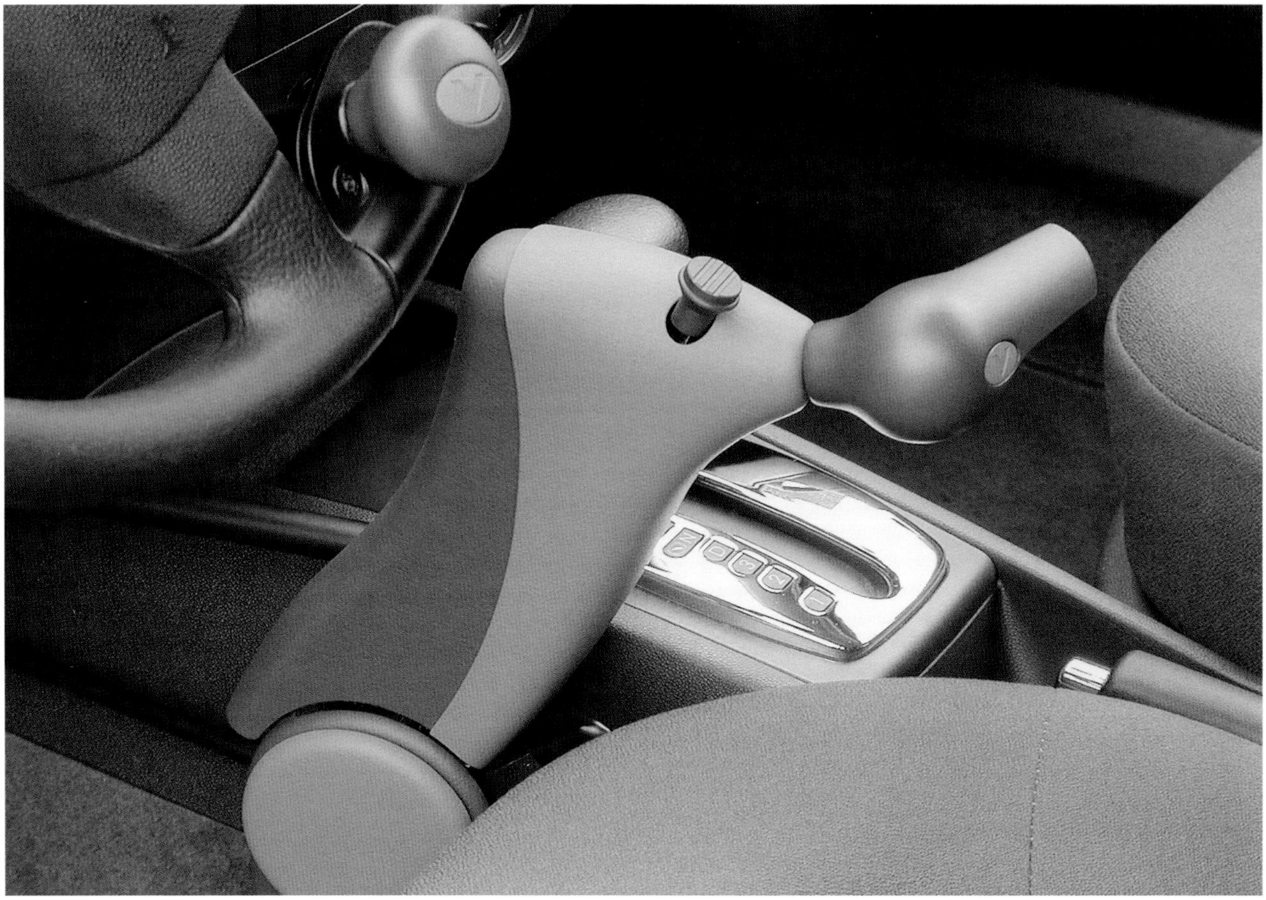

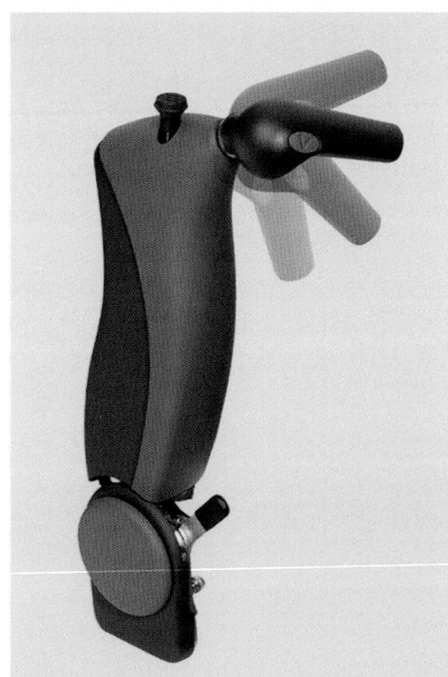

Hersteller ***Manufacturer***	**Vertrieb** ***Distributor***	**Design** ***Designer***	
Veigel GmbH + Co. KG	Veigel GmbH + Co. KG	Thamm Design	Reform Design
Lindenstraße 9–11	Lindenstraße 9–11	Hauptstraße 48	Industriestraße 25
D-74653 Künzelsau	D-74653 Künzelsau	D-74595 Langenburg	D-70565 Stuttgart
Tel. +49 79 40 91 30 17	Tel. +49 79 40 91 30 17	Tel. +49 79 05 51 22	Tel. +49 7 11 78 29 53 50
www.veigel-automotive.de	www.veigel-automotive.de	thamm-design@t-online.de	www.reform-design.de

Handbedienung für Bremse und Gas „Bruhn"
manual brake and accelerator control

Die Handbedienung ermöglicht beinbehinderten Menschen, selbst Auto zu fahren. Die bewährte sichere und solide Mechanik des Vorgängers wurde bei dem neuen Modell beibehalten. Am Markt übliche Geräte haben meist nur eine teilweise oder gar nicht verkleidete Mechanik, die negativ auffällt und ein Verletzungsrisiko birgt. Die Vollverkleidung bietet nun optimalen Schutz vor Verletzungen und den gefürchteten Druckstellen an den Beinen. Die neue Handbedienung fügt sich durch das zeitgemäße, schöne Design und die Farbgebung harmonisch in das Interieur moderner Fahrzeuge ein. Der ergonomisch verbesserte Handgriff macht zusammen mit dem zusätzlich individuell verstellbaren Winkel des Gasgriffes das Gasgeben noch leichter und ermüdungsfreier. Die Handbedienung wird im Bereich der Mittelkonsole angebracht und durch ein Gestänge mit dem Brems- und Gaspedal verbunden. Durch einfaches Entriegeln kann die Handbedienung umgeklappt werden, so daß auch Nichtbehinderte mit dem gleichen Auto problemlos fahren können.

This manual control allows people whose leg movements are impaired to drive a car. The predecessor model's tried and tested, safe and robust mechanism has been kept in the new model. In the appliances generally available on the market, the mechanism is usually only partly concealed, if at all, and this makes a negative impression as well as being a potential source of injury. The complete sheathing provides the best possible protection against injury and the dreaded bruises on the legs. With its contemporary, beautiful design and colour, the new manual control harmonizes well with the interior fittings of modern vehicles. Together with the adjustable angle of the accelerator handle, the ergonomically improved handle makes it easier and less fatiguing to accelerate. The manual control is fixed in the area of the centre console and connected by a rod to the brake and accelerator pedals. A simple release mechanism allows the manual control to be folded down, allowing people who are not handicapped to drive the same car without difficulty.

Jury ***Judges panel***

Schönes Design, läßt sich perfekt in die Hightech-Welt der Automobilindustrie integrieren, hat keinen Hauch von Behinderung. Stellt keinen Fremdkörper innerhalb dieser Hightech-Technologie dar. Ist neutral, kann dadurch auch in verschiedene Autotypen gut integriert werden.

Beautiful design, can be perfectly integrated in the high-tech world of the automotive industry, not a hint of handicap. There is nothing alien about it in the context of high-tech. Its neutral appearance allows it to be integrated well in different types of cars.

Ferrari

Focus in Gold

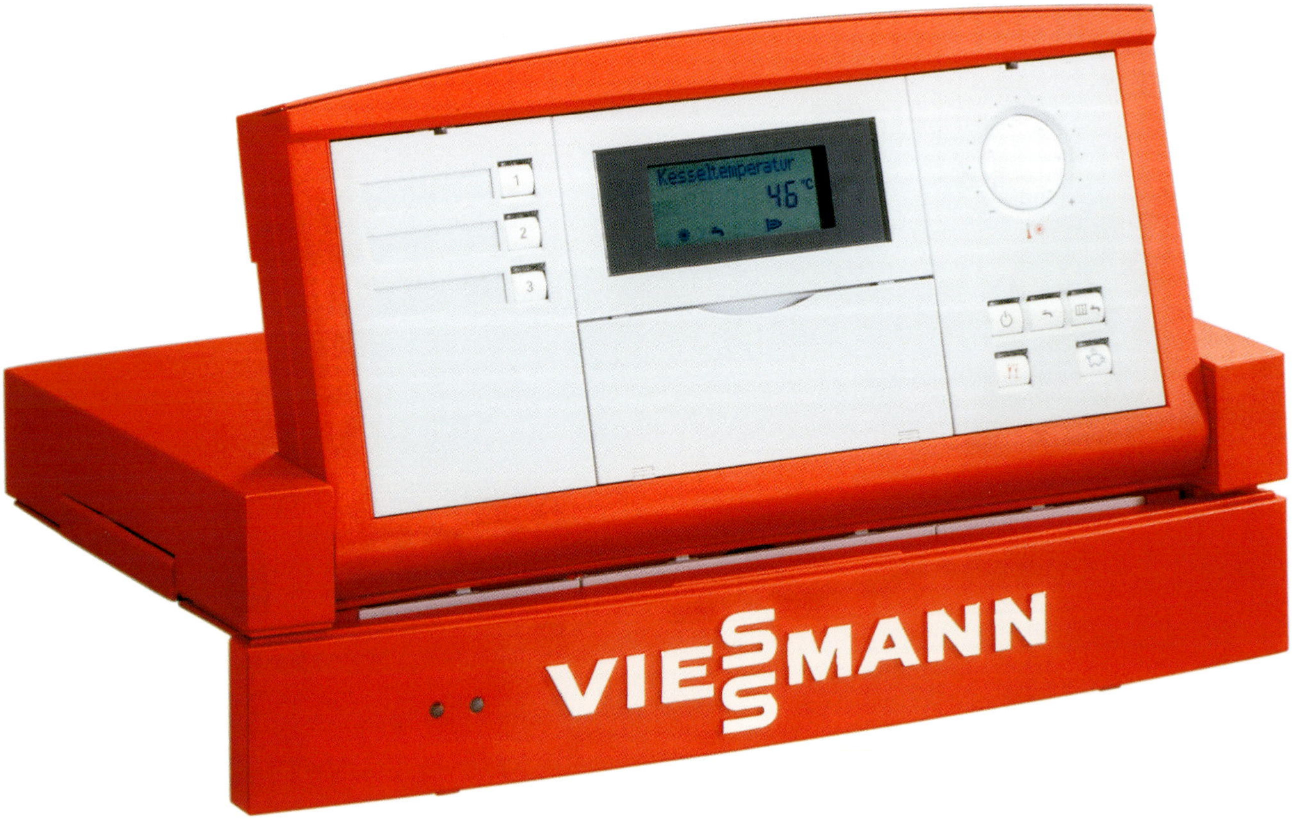

Hersteller *Manufacturer*	Vertrieb *Distributor*	Design *Designer*
Viessmann Werke D-35107 Allendorf Tel. +49 64 52 70 23 31 sgl@viessmann.com www.viessmann.com	Viessmann Werke D-35107 Allendorf Tel. +49 64 52 70 23 31 sgl@viessmann.com www.viessmann.com	Phoenix Product Design Kölner Straße 16 D-70376 Stuttgart Tel. +49 7 11 95 59 76 11 Fax +49 7 11 55 93 92

Heizkreisregelung „Vitotronic 300" *heating circuit control*

Das kommunikationsfähige, digitale Regelungssystem „Vitotronic", beinhaltet alle Funktionen. Seine Bedienelemente wurden nach neuesten ergonomischen Erkenntnissen entwickelt und anwenderfreundlich gestaltet. Das große, beleuchtete und übersichtliche Display zeigt alles in Klartext an. Die einzelnen Tasten erklären sich dank der Piktogramme selbst. So kennzeichnen etwa Sektgläser die Partytaste, mit der die Nachtabsenkung aufgehoben wird. Es gibt drei klar strukturierte Bedienebenen: von der Bedienebene für den direkten Nutzer bis zur Codierebene, die nur dem Heizungsfachmann vorbehalten ist. Viele Bedürfnisse sind von vornherein bereits erfüllt, wie zum Beispiel die automatische Sommer-/Winterzeitumstellung. Der kluge elektronische Kopf für wirtschaftliches Energie-Management erledigt viele Aufgaben, die helfen, Geld zu sparen. So ist „Vitotronic" durch Sensoren ständig über die Außentemperatur informiert und kann nach Bedarf die Temperatur steuern. Außerdem sorgt ein spezielles Ferienprogramm dafür, daß nur soviel geheizt wird wie nötig, die Wohnung nach Rückkehr aus dem Urlaub wieder wohlig warm ist und auch genügend warmes Wasser vorhanden ist.

"Vitotronic", a digital control system that can communicate contains all functions. Its control elements have been designed with the latest ergonomic findings in mind and are designed in a user-friendly way. Everything shown on the large, illuminated, clear display is in non-coded text. Thanks to the pictograms, the individual buttons are self-explanatory. For example, champagne glasses have been chosen as the symbol for the party button, which overrides the night-time cooling-down phase. There are three clearly structured operator levels: from the operator level for the actual user to the coding level reserved for the heating engineer only. Many needs have been fulfilled in advance, such as the automatic changeover to summer/winter time. The clever electronic "brain" for economical energy management performs many tasks that help save money. Sensors constantly keep "Vitotronic" informed about outside temperature, and can control temperature if necessary. In addition, a special holiday programme ensures that the heating is on only as much as necessary and that the house is nice and warm again, with plenty of hot water, when people return from holiday.

Jury ***Judges panel***

Design im Keller heißt, Heizungsingenieure beginnen, sich mit der Ästhetik von Geräten zu beschäftigen. Das freut uns sehr, daß da etwas passiert. Schönes Bedienungselement, das man mit einem Handgriff verstauen kann. Sparsame Farbgebung, einfache grafische Anordnung, hohe Technologie. Wichtig auch das Plazieren des Logos.

Design in the basement means that heating engineers are beginning to concern themselves with the aesthetics of appliances. We are very pleased to see this happening. An attractive control unit, which can be folded away in one easy movement. The colour scheme is unpretentious, the graphic arrangement is simple, the technology advanced. The logo's place is also important.

Hersteller ***Manufacturer***	**Vertrieb** ***Distributor***	**Design** ***Designer***
Merten GmbH & Co. KG Fritz-Kotz-Straße 8 D-51674 Wiehl Tel. +49 22 61 7 02 01 www.merten.de	Merten GmbH & Co. KG Fritz-Kotz-Straße 8 D-51674 Wiehl Tel. +49 22 61 7 02 01 www.merten.de	Nicholas Grimshaw & Partners Industrial Design 1 Conway Street, Fitzroy Square London W1P 6LR Tel. +44 20 72 91 41 41

Schalterprogramm „System M" *range of switches*

Bei „System M" verbinden sich eine schlichte, zeitlose, klare Formensprache und dezente Farben mit zuverlässiger Technik. Das Prinzip heißt: ein Einsatz – vier Rahmen. Der geradlinige Basisschalter M-Smart, der harmonisch geschwungene M-Arc oder der auf das Wesentliche reduzierte M-Plan lassen sich in Ein- oder Mehrfachausführung mit der einheitlichen Wippe kombinieren. Von letzterem Modell gibt es eine superflache dezente Variante M-Plan II – eine Weltneuheit –, die versenkt und flächenbündig zur Wand in fast alle Wandmaterialien eingebaut wird. Für alle vier Rahmenformen sind die Einsätze gleich, so daß die Rahmen einfach ausgetauscht werden können: vom Tast-Sensor über Jalousiesteuerungen bis zum Anzeigendisplay für den Instabus EIB für die intelligente Gebäudetechnik. Damit erfüllt „System M" auch von der technischen Seite hohe Ansprüche. Das System ist flexibel und paßt sich Änderungswünschen problemlos an. Passend zu moderner Architektur werden folgende Farben angeboten: polarweiß, aluminium, anthrazit. Das Material ist uv-stabiles Thermoplast in solider Verarbeitung, das Langlebigkeit ohne Verfärbung garantiert.

"System M" combines plain, timeless and clear formal language and discreet colours with reliable technology. The principle is one of "one insert – four frames". The plain M-Smart basic switch, the harmonically curved M-Arc, or the M-Plan which has been reduced to the bare essentials can be combined singly or in multiple units with the uniform rocker switch. The latter model is also available in the ultra-flat, discreet M-Plan II variant – a world first –, which can be installed flush in walls made of practically any material. The inserts are the same for all four frame types, allowing the frames to be simply exchanged: from the tactile sensor via window-blind controls to the display for the EIB Instabus for intelligent building installations. Technically speaking, therefore, "System M" also satisfies high standards. The system is flexible and adapts to desired modifications without difficulty. Compatible with modern architecture, the range is available in the following colours: polar white, aluminium, anthracite. They are made of UV-stable thermoplast with a robust finish, which guarantees longevity without discoloration.

Jury ***Judges panel***

Eine einfache und sinnvolle Möglichkeit in intelligentes Wohnen vorzudringen, einfache Umrüstung und angepaßt an die vertraute Wohnwelt.

A simple and sensible way of entering the sphere of intelligent housing, simple to convert, compatible with the familiar living environment.

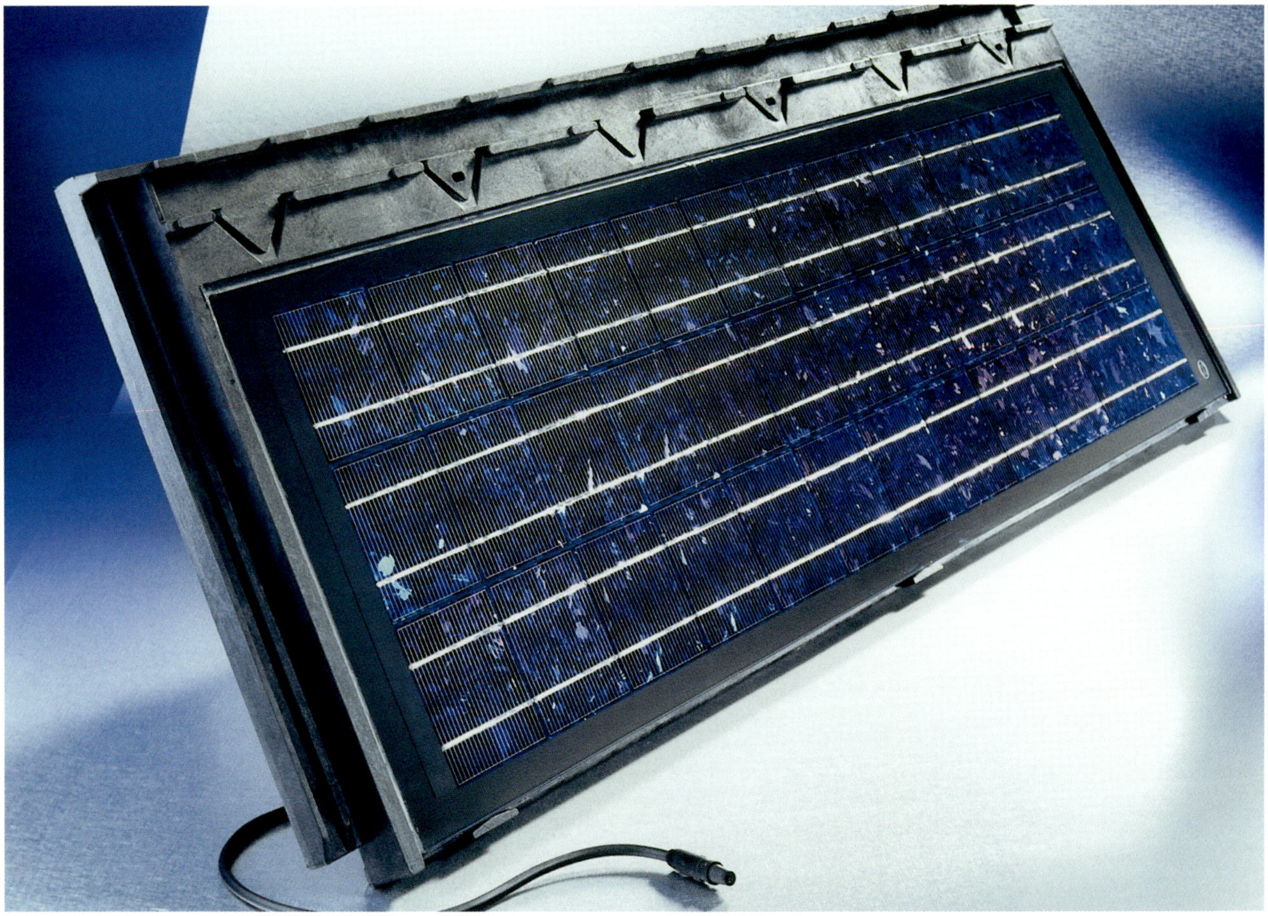

Hersteller ***Manufacturer***

Dachziegelwerke Pfleiderer
GmbH & Co. KG
Marbacher Straße 4
D-71364 Winnenden
Tel. +49 71 95 69 60
www.pfleiderer-dach.de

Vertrieb ***Distributor***

Dachziegelwerke Pfleiderer
GmbH & Co. KG
Marbacher Straße 4
D-71364 Winnenden
Tel. +49 71 95 69 60
www.pfleiderer-dach.de

Design ***Designer***

Olaf Tinzmann
Werksdesign

Tondachziegel „Terra Piatta®" *clay roof tile*
Photovoltaiksystem „Terra Piatta® – Solar" *photovoltaic system*

Der Tondachziegel „Terra Piatta®" eignet sich dank seiner schlichten Eleganz und seiner geometrischen, klar gegliederten Form für sämtliche Häuservarianten. Die Struktur des Tondachziegels wird durch seine glatte, kratzfeste Oberfläche geprägt. Die kleinen Einkerbungen am oberen Ziegelrand setzen einen unkonventionellen optischen Akzent. Zusammen mit dem leicht abgerundeten Ziegelende sichern sie gleichzeitig einen sehr guten Abfluß von Regenwasser. Dank doppelter Kopf- und Seitenverfalzungen lassen sie sich leicht verlegen und sind regensicher. Das innovative Photovoltaiksystem, das sich mit dem qualitativ hochwertigen Ziegel verbindet, ist das erste Photovoltaiksystem, das eine ästhetische sowie anwenderfreundliche Integration in die Dacharchitektur ermöglicht. Damit hebt es sich deutlich von den bisherigen, aufgeständerten Modulen ab. Der Kunststoffrahmen wurde formgleich zur Verfalzung des Ziegels konstruiert, wodurch er besonders einfach installiert werden kann und hohe Funktionalität und maximale Regensicherheit garantiert. „Terra Piatta® – Solar" ist eine attraktive Alternative, umweltschonend Strom aus Sonnenlicht zu erzeugen.

Thanks to its modest elegance and geometric, clearly structured form, the "Terra Piatta®" clay roof tile is suitable for all types of houses. The structure of the tile is characterized by a smooth, non-scratch surface. The small notches on the top edge of the tile are an unconventional optical accent. Together with the slightly rounded tile edge, they also ensure that rain water runs off very well. Double grooves at the top and sides allow them to be laid easily, as well as making them rainproof. The innovative photovoltaic system, which is combined with this high-quality tile, is the first photovoltaic system to allow integration in roof architecture that is both aesthetically pleasing and user-friendly. It clearly stands out from the modules on their stands that have been usual so far. The plastic frame has been designed to fit into the groove on the tile, making it particularly easy to install, and guaranteeing a high level of functionality and maximum protection against rain. "Terra Piatta® – Solar" is an attractive alternative for the non-polluting generation of power from sunlight.

Jury ***Judges panel***

Sinnvolle Technik in der Architektur versteckt. Angenehme Integration der Solartechnik in das Keramikdach.

Practical technology concealed in architecture. Pleasant integration of solar technology in a tiled roof.

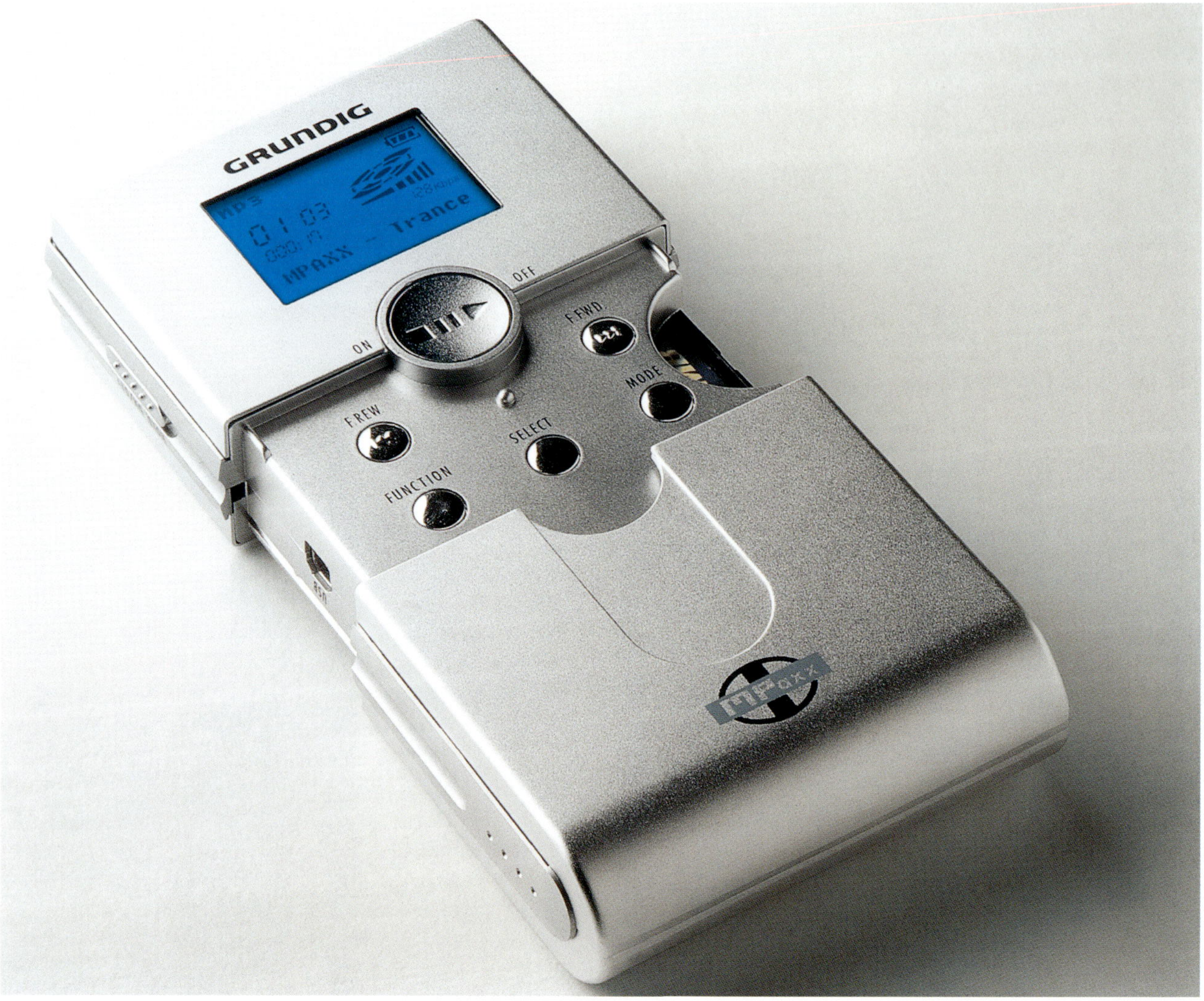
GRUNDIG
MPAXX - Trance
ON
OFF
F.FWD.
MODE
F.REW.
SELECT
FUNCTION

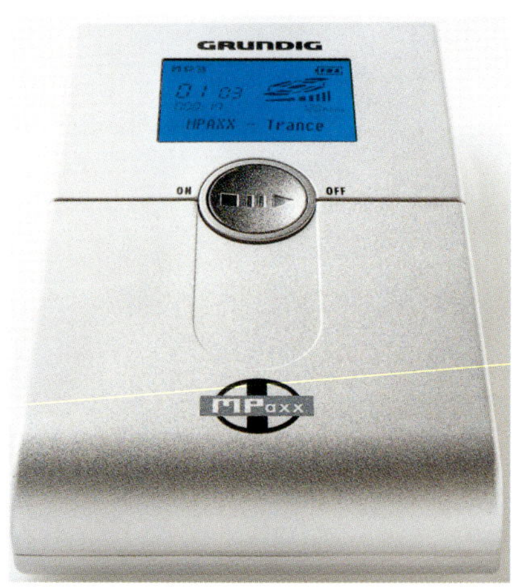
GRUNDIG
MPAXX - Trance
ON
OFF

Hersteller ***Manufacturer***

Grundig AG
Beuthener Straße 41
D-90471 Nürnberg
Tel. +49 9 11 7 03 72 53
www.grundig.com

Vertrieb ***Distributor***

Grundig AG
Beuthener Straße 41
D-90471 Nürnberg
Tel. +49 9 11 7 03 72 53
www.grundig.com

Design ***Designer***

Werner Engels
Grundig Produkt Design

Digitaler Audio Player „MPaxx SP 4101" *digital audio player*

Der „MPaxx SP 4101" ist ein elegantes Musikabspielgerät für Trendsetter – viel Musik auf kleinstem Raum. Mit ihm kann man unterwegs Musik aus dem Internet hören. Er ist ein portabler Audio Player für alle gängigen, digitalen Audio-Kompressionsverfahren für Musik im Internet, wie MP 3, WMA und AAC-Intertrust. Er kann diese Musik – oder auch Musikdaten von der CD – über den PC entweder auf kleine Multi-Media-Cards / Secure Digital Cards oder auf den integrierten Flash-Speicher downloaden und wiedergeben. So kann man sich digitale Musikprogramme individuell zusammstellen und unterwegs hören. Denn „MPaxx SP 4101" ist nicht nur klein, handlich und leicht – er wiegt gerade mal 82 Gramm – sondern auch leistungsstark und unempfindlich gegen Erschütterungen. Er liest die Musikdaten ohne jede mechanische Bewegung direkt vom Chip der Speicherkarte – das Gerät besitzt zwei Einschübe für Speicherkarten – beziehungsweise aus dem integrierten Speicher. Um an die Bedienelemente zu kommen, wird das Gehäuse dank einer Schiebemechanik einfach auf- und zugeschoben. Das Display ist beleuchtbar. Das Gehäuse ist aluminiumfarben und unterstreicht das sachliche technische Design.

The "MPaxx SP 4101" is an elegant music player for trendsetters – a lot of music in the smallest of spaces. It can be used on the move to listen to music from the internet. It is a portable audio player for all the current digital audio compression methods for music in the internet, such as MP 3, WMA and AAC-Intertrust. Via a PC, it can download this music – or also music data from a CD – either onto small MultiMediaCards / SecureDigitalCards or onto its integrated flash card, and play it. This allows digital music programmes to be put together and then listened to while on the move. "MPaxx SP 4101" is not only small, handy and light – weighing just 82 grams – but is also powerful and insensitive to shock. It reads music data without any mechanical movement direct from a chip or memory card – the device has two slots for flash cards – or from its integrated memory. To access the controls, the housing is simply slid back by means of a sliding mechanism. The display can be illuminated. The housing's aluminium colour emphasizes its sober, technical design.

Jury ***Judges panel***

Hightech-Technologie und -Design, funktional überzeugend. Sympathisches Maß, einfach in der Tasche zu tragen. Versteckte Bedienung. Interessanter Schiebemechanismus. Bei der Vielfalt von Technologie Überraschungseffekt beim Betätigen: ist es ein Fotoapparat oder Mini-Disc-Player? Das heißt die Funktionen sind fast austauschbar geworden.

High technology and high-tech design, functionally convincing. Nice size, simple to carry in your pocket. Concealed controls. Interesting sliding mechanism. With the wealth of technology, an effect of surprise when operating it: is it a camera or a mini-disc player? In other words, functions have become almost interchangeable.

EXIT
e-Mail - New
to INJ
Subj. Header 01
Die amerikanische Gesu
ndheitsbehörde FDA hat
erstmals ein Verfahren z
ur Kontrolle des Gesund
heitszustands von Patien
ten per Internet zugelas
sen.
Enter
Back
Menu
'01
21.Dez.
pm
00:00
FILEWALKER

Hersteller ***Manufacturer***	**Vertrieb** ***Distributor***	**Design** ***Designer***
Invair Technologies AG	Birkhold KG	Steffen Schmidt
Wollgrasweg 49	Vogelsangstraße 16a	Luruper Weg 19
D-70599 Stuttgart	D-70176 Stuttgart	D-20257 Hamburg
Tel. +49 1 60 7 14 14 83	Tel. +49 7 11 66 98 50	Tel. +49 1 60 7 14 14 83
www.invair.com	Fax +49 7 11 6 69 85 90	Fax +49 7 11 4 51 01 75 19

Personal Digital Assistant „Filewalker" *PDA*

Der „Filewalker" bietet echte Mobilität und hohen Nutzen auf kleinstem Raum. Er ist nicht nur leicht, klein, handlich, paßt in die Hosentasche – denn er hat das Format einer Scheckkarte und wiegt nur ganze 98 Gramm –, sondern auch einfach zu bedienen. Neben einem Adreßbuch besitzt er einen Kalender zur Terminverwaltung mit Alarmfunktion, eine Aufgabenliste, ein Notizbuch und eine Weltzeituhr. Man kann ihn als Stoppuhr und Taschenrechner nutzen, mit ihm Emails übers Handy oder den PC empfangen und versenden. Der „Filewalker" zeichnet sich durch hohe Ergonomie und intuitive und selbsterklärende Benutzerführung aus. Im Vergleich zu anderen Geräten ist hier neu, daß man mit nur einer Hand, die in einer natürlichen Stellung gehalten wird, die einzelnen Anwendungen über Scrollband und die drei seitlichen Tasten vollständig bedient sowie Text eingibt. Dabei wird das hochauflösende Display mit blauer Hintergrundbeleuchtung maximal ausgenutzt. Mit den drei Tasten über dem Display wird das Gerät bedient und die verschiedenen Anwendungen ausgewählt. Der „Filewalker" besitzt ein stabiles Metallgehäuse aus glasperlgestrahltem und eloxiertem Aluminium.

"Filewalker" provides true mobility and many benefits in the smallest of spaces. Not only is it light, small, handy and pocket-sized – for it is the same size as a cheque card, and weighs just 98 grams –, it is also simple to operate. Apart from an address book, it has an appointments diary with an alarm function, a to do list, a notebook and a world-time clock. It can be used as a stopwatch and pocket calculator, to receive and send emails via a mobile telephone or PC. "Filewalker's" ergonomic design is distinctive, as are the intuitive and self-explanatory user controls. Compared with other devices, one new feature is that one hand is sufficient to operate all the applications, using the scroll key and three keys on the side, and to enter text – all this while keeping the hand in a natural position. The high-resolution display with its blue backlighting is exploited to the maximum. The device is operated and the various applications selected by the three keys above the display. "Filewalker" has a robust metal housing of glass-pearl blasted and anodized aluminium.

Jury ***Judges panel***

Äußerst kleines Format. Wermutstropfen ist sicher die Rechtshändigkeit. Überzeugend das Dreitastensystem, bei dem man schneller zugreifen kann als mit dem Stift. Kompletter Einhänder, die anderen sind immer noch Zweihänder. Die Gehäusebefestigung ist sehr filigran und gut gelöst ohne Schrauben, durch Klicken. Die Bildschirmgröße im Verhältnis zur Größe des Gerätes ist optimal.

Extremely small format. Of course, it is bitter that it is only for right-handers. The three-button system is convincing, allowing faster access than with a stylus. Completely one-handed device, the others still require both hands. The housing is very filigree, well worked out, clipped into place without screws. Optimum ratio of screen size to device size.

THIN STATION

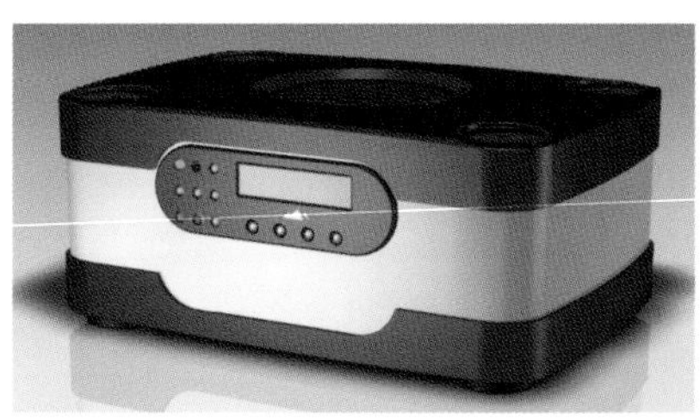

Hersteller *Manufacturer*	**Vertrieb** *Distributor*	**Design** *Designer*
Pyramid Computer Systeme GmbH Bötzinger Straße 60 D-79111 Freiburg Tel. +49 7 61 4 51 47 11 www.pyramid.de	Pyramid Computer Systeme GmbH Bötzinger Straße 60 D-79111 Freiburg Tel. +49 7 61 4 51 47 11 www.pyramid.de	Formgestaltung Schnell-Waltenberger Sponheimstraße 22 D-75177 Pforzheim Tel./Fax +49 72 31 3 33 64

Internet Server „BenHur cigarbox™" *internet server*

Die „BenHur cigarbox™" ist ein Miniatur-Internetserver, der auf der Basis der bereits seit langem im Industriebereich erfolgreich eingesetzten BenHur-Lösung nun erstmals als intelligente Infrastrukturkomponente ein Stück des WWW in den privaten Haushalt bringt. Der Server ist gewissermaßen wie ein Router die Verbindung zur Außenwelt, gleichzeitig jedoch auch Web Stellvertreter, Speicher und Kontrolleur. Die „cigarbox™" ist ein völlig neues Gehäuse-Konzept für Miniatur-Computer in unterschiedlichsten Anwendungsbereichen: viel Funktion auf kleinstem Raum zum geringstmöglichen Preis. Das universelle Computergehäuse hat die Größe einer halben DIN A4-Seite. Die Basis bilden ein umschlagsgleiches Kunststoffteil und ein auf das jeweils verwendete Motherboard abgestimmtes Blechstanzteil. Benötigt man eine größere Höhe, wenn Erweiterungskarten eingesetzt werden, so erhält man diese mit einem umlaufenden Blech. Das Gerät läßt sich nicht nur im Querformat hinstellen sondern auch als Thin Client im Hochformat mit zusätzlichem Alufuß einsetzen oder mit einem Befestigungsset an Wand, Decke, unter dem Tisch oder im Auto anbringen.

The "BenHur cigarbox™" is a miniature internet server. Based on the BenHur solution that has long been successfully applied in industry, this intelligent infrastructure component is now available for the first time as a way of bringing the web to the private household. To a certain extent, the server is like a router, providing a connection to the outside world, but at the same time it is a web representative, memory and supervisory facility. The "cigarbox™" is a completely new chassis concept for miniature computers in the widest variety of application environments: a lot of functions in the smallest possible space at the lowest possible price. The universal computer chassis is the size of half a sheet of DIN A4 paper. The basic elements are two identical plastic elements and a pressed metal element compatible with the motherboard layout selected. If more height is needed, for example for expansion boards, a new sheet metal element provides this. The appliance can either be laid flat or, with the help of an additional aluminium stand, be stood on end as a thin client, or be fixed to the wall, to the ceiling, under a desk or in the car with the help of a fastening kit.

Jury ***Judges panel***

Sehr kompakt, schöne Oberflächengravuren. Grafik, Material und Farbe laufen homogen zusammen. Robust wie eine Blackbox. Gestalterisch bricht das Produkt aus dem Hightech-Silberoberflächen-Design aus. Sehr schöne Haptik. Bei solchen Geräten würde die Haptik auch bedeuten, es ist nicht kühl, es ist warm und durch die Profilierung schwitzt man nicht an den Fingern.

Very compact, beautiful surface engraving. Graphics, material and colour form a homogenous whole. As robust as a black box. Design-wise, the product gets away from high-tech silver-surface design. Very pleasant to the touch. In devices such as these, quality of touch also means being warm rather than cool, your hands do not sweat when touching it.

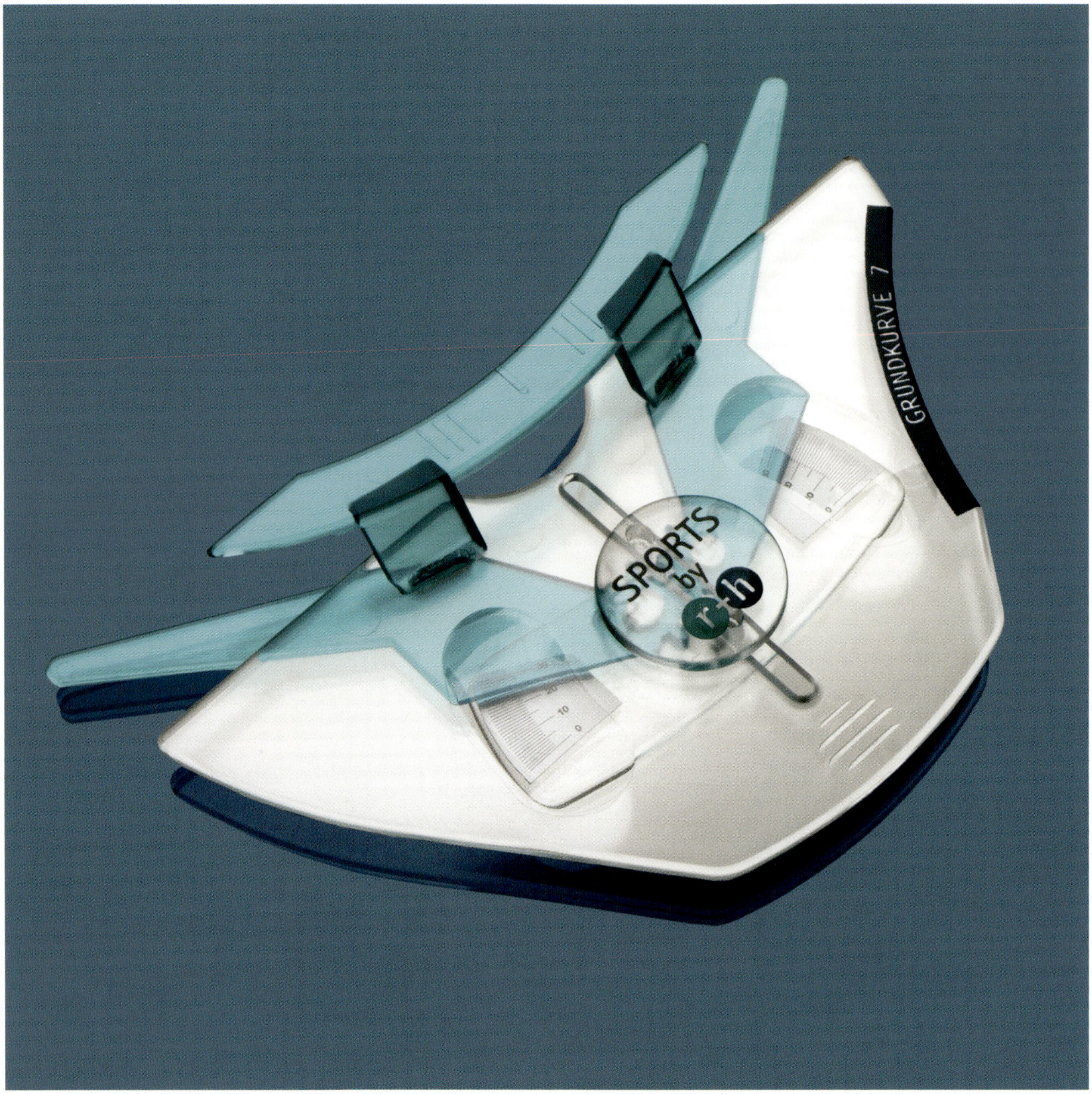
GRUNDKURVE 7
SPORTS
by
r-h

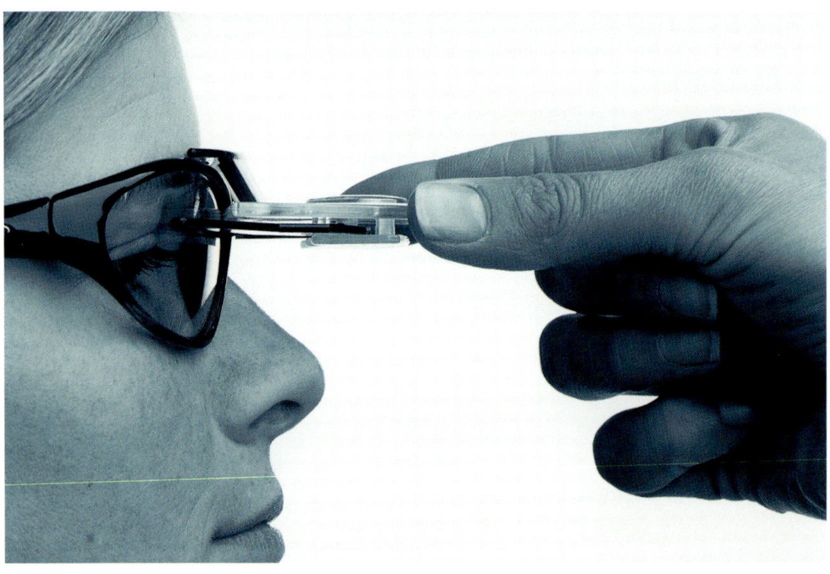

Hersteller *Manufacturer*

Rupp + Hubrach Optik GmbH
Von-Ketteler-Straße 1
D-96050 Bamberg
Tel. +49 9 51 1 86 13 31
www.rh-brillenglas.de

Vertrieb *Distributor*

Rupp + Hubrach Optik GmbH
Von-Ketteler-Straße 1
D-96050 Bamberg
Tel. +49 9 51 1 86 13 31
www.rh-brillenglas.de

Design *Designer*

Lutz Gathmann
Nagelsweg 41
D-40474 Düsseldorf
Tel. +49 2 11 45 29 19
www.lutz-gathmann.de

Meßgerät „Sports" *measuring device*

Es gibt immer mehr und differenziertere Brillen für die unterschiedlichsten Sportarten. Brillenträger haben damit, wenn sie beim Sport nicht auf Kontaktlinsen ausweichen wollen, ein Problem. Die Sportfassung ist stärker durchgebogen, so daß die Brille fast parallel zur Kopfform verläuft und die Brillengläser schräg vor dem Auge liegen. Folglich steht die Sehachse nicht senkrecht auf der Brillenglasfläche, das Blickfeld ist verzerrt. Durch eine revolutionäre neue Brillenglastechnik ist es gelungen, den Winkel zwischen Glas und Durchblicksebene sowie die stärkere Krümmung der Gläser bei der Produktion des optischen Brillenglases zu kompensieren. Dazu muß aber der Winkel bei getragener Brille am Kopf des Sportbrillenträgers exakt bestimmt werden. Dafür wurde das Meßgerät „Sports" entwickelt. Mit seiner Hilfe ist es nun erstmals möglich, auch Radsport- oder Skibrillen für Brillenträger anzupassen. Das Meßgerät ist nicht nur hart, bruchsicher, flexibel und dabei federleicht, sondern auch einfach und praktisch zu bedienen. Dazu trägt bei, daß die einzelnen Elemente und Funktionen durch unterschiedliche Farbbereiche logisch und klar getrennt worden sind.

Ever new types of glasses are being devised for the widest variety of sports. If they do not wish to wear contact lenses for sport, this presents a problem for people who wear glasses. Sports glasses are more strongly curved, which means that the glasses run almost parallel to the shape of the wearer's head, and the lenses are at an angle to the eyes. Consequently, the line of vision is not 90° to the surface of the lenses, and the field of vision is distorted. A revolutionary new lens technology now allows the angle between lens and line of vision, as well as the greater curvature of the lenses, to be compensated during the manufacture of lenses. For this to be possible, however, the angle must be determined exactly when the glasses are actually being worn. The "Sports" measuring device has been developed for this purpose. Using this device, it is now possible for the first time to fit spectacle-wearers with glasses for cycling or skiing. The measuring device is not only hard, unbreakable, flexible and ultra-light, it is also simple and practical to use. This is helped by using different colours to separate the individual elements and functions logically and clearly.

Jury *Judges panel*

Ein kleines, handliches Meßgerät, das das Anpassen von Sportbrillen schon fast in eine medizinische Ecke rückt. Sympathisch, daß der Maschinencharakter entfällt, daß es in gewissem Sinne ein Hand-Augen-Gerät wird. Schöne übersichtliche Grafik plus Farbspiel und Mechanik. Schaut im positiven Sinne handgemacht aus, not over-engineered.

A small, handy measuring device, which almost propels the fitting of sports glasses into the medical sphere. It is good to see that the machine character has been abandoned, and that it has become a hand-eye device, as it were. Good, clear graphics, combined with colour coding and mechanical elements. A hand-made appearance in the positive sense, not over-engineered.

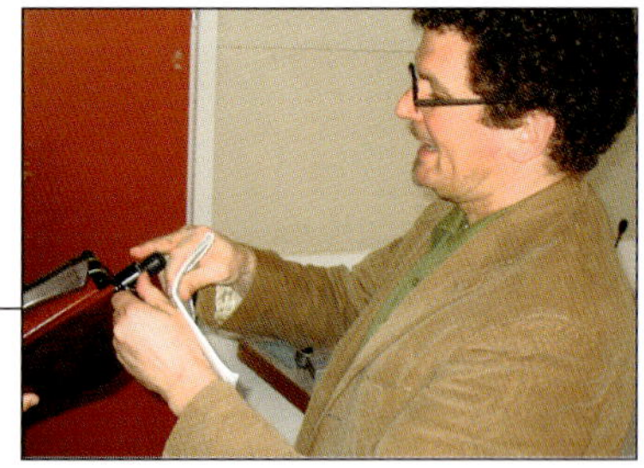

Fritz Frenkler

Designer, f/p design Frankfurt
Designer, f/p design Frankfurt

Studierte Industriedesign und Psychologie in Braunschweig. Ab 1984 arbeitete er als Designer bei frogdesign in Altensteig und in Campbell/USA. Von 1986 bis 1991 hatte er die Geschäftsführung von frogdesign in Asien, bevor er 1992 nach Deutschland zurückging und zur wiege Entwicklungsgesellschaft in Bad Münder wechselte. Von 1999 bis 2001 war Fritz Frenkler Geschäftsführer der Deutschen Bahn Medien GmbH. Anschließend eröffnete er mit Anette Ponholzer das Büro f/p design in Frankfurt. Fritz Frenkler gestaltete Produkte für international tätige Firmen verschiedenster Branchen: Wilkhahn, Kukuyo, Olympus zählen zu seinen Kunden. Mit Ausstellungssystemen und -pavillons auf der Expo 2000 in Hannover hatte Fritz Frenkler genauso Erfolg wie mit anderen Gestaltungskonzepten für den öffentlichen Raum, zum Beispiel für die Verkehrsbetriebe ÜSTRA in Hannover.

Studied Industrial Design and Psychology in Brunswick. From 1984, he worked as a designer at frogdesign in Altensteig and in Campbell/USA. From 1986 to 1991, he was the managing director of frogdesign in Asia, returning to Germany in 1992 to take up a position with wiege Entwicklungsgesellschaft in Bad Münder. From 1999 to 2001, Fritz Frenkler was managing director of Deutsche Bahn Medien GmbH. Subsequently, he and Anette Ponholzer opened the f/p design bureau in Frankfurt. Fritz Frenkler has designed products for globally operating companies in the widest variety of industries: Wilkhahn, Kukuyo and Olympus are among his clients. Fritz Frenkler's successes include exhibition systems and pavilions at the Expo 2000 in Hannover and other design concepts for the public sphere, such as designs for ÜSTRA transport services in Hannover.

Birgit Mager

Organisations- und Personalentwicklerin, Professorin für Service Design an der Fachhochschule Köln
Organization and Human Resources Developer, Professor of Service Design at the Cologne Fachhochschule (College of Commerce and Industry)

Studium der morphologischen Psychologie an der Universität zu Köln. Danach war sie mehrere Jahre als Organisations- und Personalentwicklerin im Bereich Human Resources Development bei Hewlett Packard in Böblingen tätig. In einem großen Restrukturierungsprozeß begleitete sie als Leiterin der Personal- und Organisationsentwicklung die Muelhens KG in Köln. Als selbstständige Beraterin spezialisierte sie sich auf die Service Industrie. Seit 1995 hat sie den europaweit noch einzigen Lehrstuhl für Service Design am Fachbereich Design der Fachhochschule Köln inne. Zahlreiche Artikel und Bücher zum Thema Service Design, Gastprofessuren in Hongkong, Chicago, Zürich, Luzern und Wuxi, China. Jurypräsidentin für den Service Design Preis Schweiz und Mitglied in vielen anderen internationalen Design Jurys. Realisation zahlreicher Seminare, Kongresse und insbesondere Ausstellungen zum Thema Service.

Studied Morphological Psychology at the University of Cologne. Subsequently several years experience as organization and human resources developer in the Human Resources Development department at Hewlett Packard in Böblingen. In her capacity as head of human resources and organizational development she advised Muelhens KG in Cologne in a major restructuring process. As a freelance consultant, she has specialized in the service industry. Since 1995, she has held the only Service Design chair in Europe, in the Design faculty of the Cologne Fachhochschule. She has not only published many articles and books on the subject of Service Design, but has also been a visiting lecturer in Hong Kong, Chicago, Zurich, Lucerne and Wuxi, China. She is the chairperson of the judges' panel for the Swiss Service Design Award and a member of several other judges panels for international design awards. She has organized several seminars, congresses and, above all, exhibitions dealing with the topic of service.

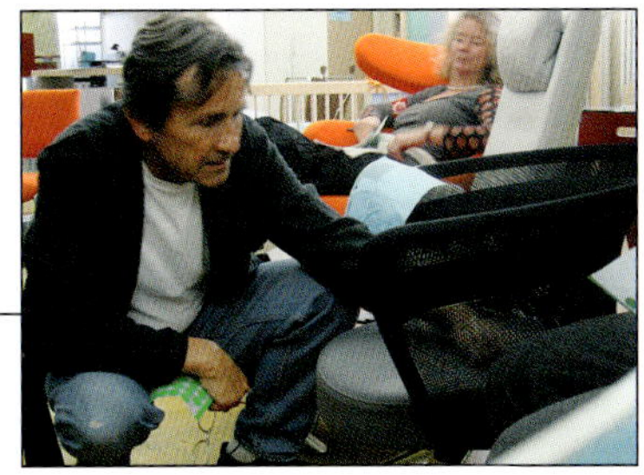

Hansjerg Maier-Aichen

Designer und Kulturberater, Stuttgart, Professor für Produktdesign an der Hochschule für Gestaltung Karlsruhe
Designer and cultural advisor, Stuttgart, Professor of Product Design at the Hochschule für Gestaltung (Design College) Karlsruhe

Studierte zunächst Innenarchitektur und danach Malerei in Deutschland, Frankreich und in den USA. Nach verschiedenen Stipendien unter anderem in den USA, Deutschland und Italien, zahlreichen Ausstellungen in Europa und Lehraufträgen an deutschen und amerikanischen Hochschulen war er von 1974 bis 1984 Kulturberater der Europäischen Union. 1983 gründete er die Designmarke Authentics, unter der Kunststoffkollektionen herausgebracht wurden. In wenigen Jahren entwickelte sich die Marke zu einem weltweiten und beispielhaften Unternehmen für anspruchsvolle Designprodukte des Alltags, die zahlreiche nationale und internationale Designauszeichnungen gewannen und in vielen Museen vertreten sind. Internationale Publikationen über das Authentics Konzept. Mitglied in Vorständen und Präsidien. Seit 2002 Professor für Produktdesign an der Hochschule für Gestaltung Karlsruhe.

Studied first Interior Design then Art in Germany, France and the USA. Following various scholarships, for example in the USA, Germany and Italy, many exhibitions in Europe and lectureships at German and American universities, he was the cultural advisor to the European Union from 1974 to 1984. In 1983, he founded the Authentics design brand, under whose name ranges of plastic products were launched. Within a few years, the brand developed to become a worldwide model company for sophisticated everyday designer products, which won many national and international design awards and are to be found in many museums. International publications on the Authentics concept. Member of advisory and governing boards. Since 2002, professor of Product Design at the Karlsruhe Hochschule für Gestaltung.

Wim Ouboter

Erfinder, Micro Mobility Systems Küsnacht
Inventor, Micro Mobility Systems Küsnacht

Der Erfinder des Kickboards, das als besonders innovatives Produkt bereits 1999 den Internationalen Designpreis Baden-Württemberg gewann, ist ein Quereinsteiger. Nach einer Banklehre in der Schweiz, einem Hochschulabschluß als Betriebsökonom und einem Marketingstudium an der Universität Boston baute er zusammen mit einem Partner in South Carolina eine Produktionsfirma für technische Textilien auf. Wieder zurück in der Schweiz war er zunächst in der Immobilienbranche und im Verwaltungsrat einer Computerhandelskette tätig. Schließlich konnte er seine Vision von Micro Mobilität in dem ersten Micro-Tretroller und dem Kickboard verwirklichen, die beide ein riesiger Erfolg wurden. 1996 gründete er Ouboter Brothers Ltd. Micro Mobility Systems. Schwerpunkt der heutigen Tätigkeit sind insbesondere das Erkennen langfristiger Trends und die Realisierung außergewöhnlicher Produkte für globale Märkte.

The inventor of the kickboard, which won the Baden-Württemberg International Design Award as early as 1999 in recognition of its highly innovative design, came to design via an unconventional career path. Following an apprenticeship in a Swiss bank, a university degree in Business Economics and a course in Marketing at the University of Boston, he and a partner established a production company for technical textiles in South Carolina. On returning to Switzerland, he first worked in the real estate business, then on the administrative board of a computer marketing chain. Finally, he was able to realize his vision of micro-mobility with the first micro-scooter and the kickboard, both of which were a huge success. In 1996 he founded Ouboter Brothers Ltd. Micro Mobility Systems. Today, the company focuses in particular on recognizing long-term trends and producing unusual products for global markets.

Barbara Schmidt

Designerin, Kahla / Thüringen Porzellan GmbH
Designer, Kahla / Thüringen Porzellan GmbH

Studierte an der Hochschule für Kunst und Design Halle / Burg Giebichenstein und der University of Art and Design in Helsinki. Im Anschluß daran war sie in der Keramikindustrie zuerst in Finnland und später in den Niederlanden beschäftigt. Seit 1991 arbeitet sie als Designerin bei der Kahla/Thüringen Porzellan GmbH. Sie zeichnet für viele moderne und innovative Produktprogramme des Unternehmens verantwortlich, die sie gestaltet hat. 1999 organisierte sie für Kahla den Designwettbewerb Survival Kit. Außerdem nahm sie an vielen Ausstellungen in verschiedenen Ländern Europas und in den USA teil und erhielt zahlreiche international renommierte Designpreise. Seit 1999 hat Barbara Schmidt zusätzlich einen Lehrauftrag an der Hochschule für Kunst und Design in Halle.

Studied at the Halle / Burg Giebichenstein Hochschule für Kunst und Design (Art and Design College) and at the University of Art and Design in Helsinki, subsequently working in the ceramics industry, first in Finland and later in the Netherlands. Since 1991, she has worked as a designer at Kahla/Thüringen Porzellan GmbH, where she has designed and been responsible for many modern and innovative product ranges. In 1999, she organized the Survival Kit design competition for Kahla. She has also taken part in many exhibitions in various European countries and the USA, and received many internationally acclaimed design awards. Since 1999, Barbara Schmidt has also been a lecturer at the Halle Hochschule für Kunst und Design.

WAS MEHR ALS EINEM NÜTZT

Im **Focus Lebensart:**

Mia Seeger Preis 2002
für junge Designerinnen und Designer, genauer Absolventen der Studiengänge Industriedesign/Produktgestaltung, Innenarchitektur/Möbeldesign, Investitionsgüter- oder Transportation Design an deutschen Hochschulen zwischen 2000 und 2002.

Gesucht sind Diplom- oder vergleichbare Entwurfsarbeiten, die sich mit unserer Art zu leben und zusammenzuleben befassen und hierfür sinnreiche, sozial nützliche Lösungen vorschlagen. Weitere Kriterien: umweltverträglich, neuartig und auf hohem gestalterischen Niveau.

1. Preis 3.500 €   2. Preis 2.500 €   3. Preis 1.500 €

Jury
Lore Elbracht, designafairs GmbH, München
Nils Holger Moormann, Moormann Möbel GmbH, Aschau i.Ch.
Tom Schönherr, Phoenix Product Design, Stuttgart
Gerhard Zambelli, ehem. Bezirksleiter der IG-Metall Baden-Württemberg
Dr. Wolfgang Berger, Design Center Stuttgart

Publikation, Preisverleihung und Präsentation zusammen mit dem Internationalen Designpreis Baden-Württemberg im Design Center Stuttgart.

ERGEBNISSE
RESULTS

Einsendeschluß
5. Juni 2002
Ausführliche Wettbewerbsunterlagen bei der
Mia Seeger Stiftung
c/o Design Center Stuttgart
Willi-Bleicher-Straße 19
70174 Stuttgart
Telefon 0711/123-27 81
Fax 0711/123-25 77
E-mail design@lgabw.de

Erster Preis (3.500 €) Drohne statt Spürhund

First prize (3,500 €) Drone instead of tracker dog

Holger Pilz, Wäschenbeuren, pilz-h@hfg-gmuend.de, Hochschule für Gestaltung FH Schwäbisch Gmünd
Stefan Schenk, Grünenbach, schuettentobel@hotmail.com, Hochschule für Gestaltung FH Schwäbisch Gmünd

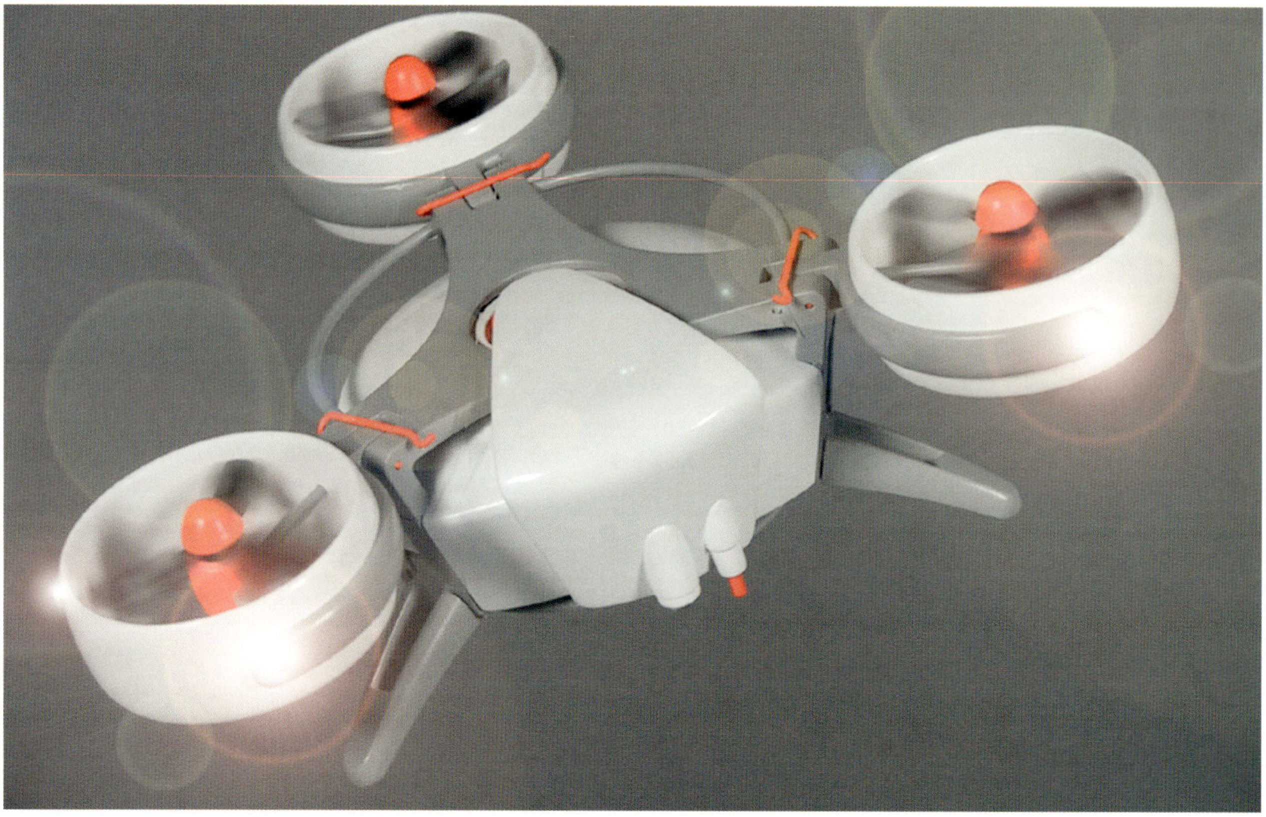

Das Gerätesystem „remote rescue" zum Auffinden von verschütteten Erdbebenopfern umfaßt ein unbemanntes, ferngesteuertes Fluggerät mit Bioradar, der Atem und Herzschlag erkennen kann, eine Handsteuerung, zwei Bildschirmkontrollgeräte und einen Transportcontainer, der vor Ort zum Arbeitsplatz für die Einsatzleitung wird.

The "remote rescue" set of appliances for finding buried earthquake victims comprises an unmanned, remote controlled flying object with BioRadar, which can sense breathing and heart beats, a hand-held control unit, two control units with VDUs and a transport container which can be used on the spot as a command centre.

Jury *Judges panel*

Dem Entwurf gingen sorgfältige Analysen von Rettungsaktionen und eingehende technologische Recherchen voraus. Mit Sensorik, Mobilität und Kommunikation sind die entscheidenden Aspekte für schnelle Ortung, risikomindernde Bergung und koordinierte Einsatzleitung erfaßt. Konsequent folgt der Entwurf dem konzeptionell vorgezeichneten Weg. Die Komponenten sind so entwickelt, daß sie sowohl einzeln im Gebrauch als auch im Zusammenspiel zu einem optimierten Rettungseinsatz beitragen. Insgesamt ist ein neuartiges, einsatzgerechtes und in vielen praktischen Details akribisch ausgearbeitetes Gerätesystem entstanden.

The design was preceded by careful analyses of rescue operations and detailed technological research. Together with sensor technology, the decisive features for locating victims, low-risk rescue work and coordinated command are mobility and communication. The design sticks rigorously to the original concept. The components have been designed to make the best possible contribution to rescue work, both when being used alone and in combination with other components. All in all, the result is an innovative, practical set of appliances, whose many useful details are evidence of great attention to detail.

Einsatzszenario: *Deployment scenario:*

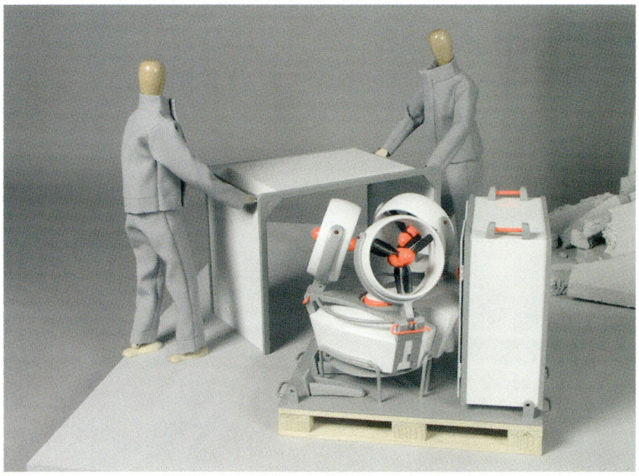

Die Ausrüstung wird am Einsatzort ausgepackt.
Unpacking the equipment at the scene of the disaster.

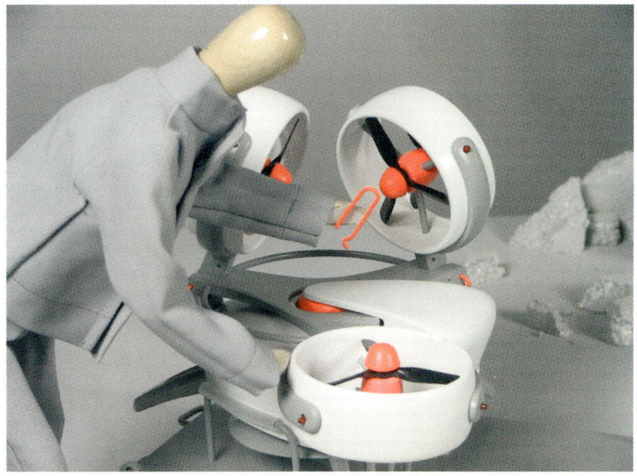

Das Fluggerät wird startklar gemacht.
The flying object is made ready for take-off.

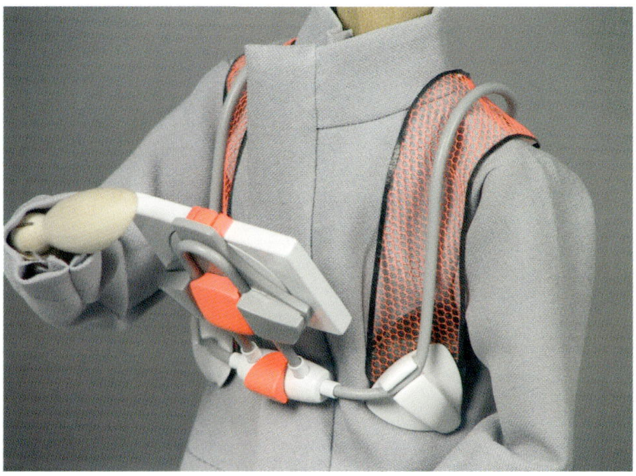

Kontrollgerät und Handsteuerung werden angelegt.
The control units with VDUs and hand-held control unit are put on.

Das Fluggerät über dem Einsatzgebiet.
The flying object in action.

Die Einsatzleitung fertigt vom Unglücksort ein Luftbild.
The rescue team puts together an aerial photograph of the scene of the disaster.

Ortungseinsatz: Navigation und Datenübermittlung.
Locating victims: navigation and data transmission.

Zweiter Preis (2.500 €) Rollstuhl vom Feinsten

Second prize (2,500 €) Deluxe wheelchair

Matthias Kolb, Pforzheim, ilo-01@gmx.net, Hochschule für Gestaltung Technik und Wirtschaft FH Pforzheim

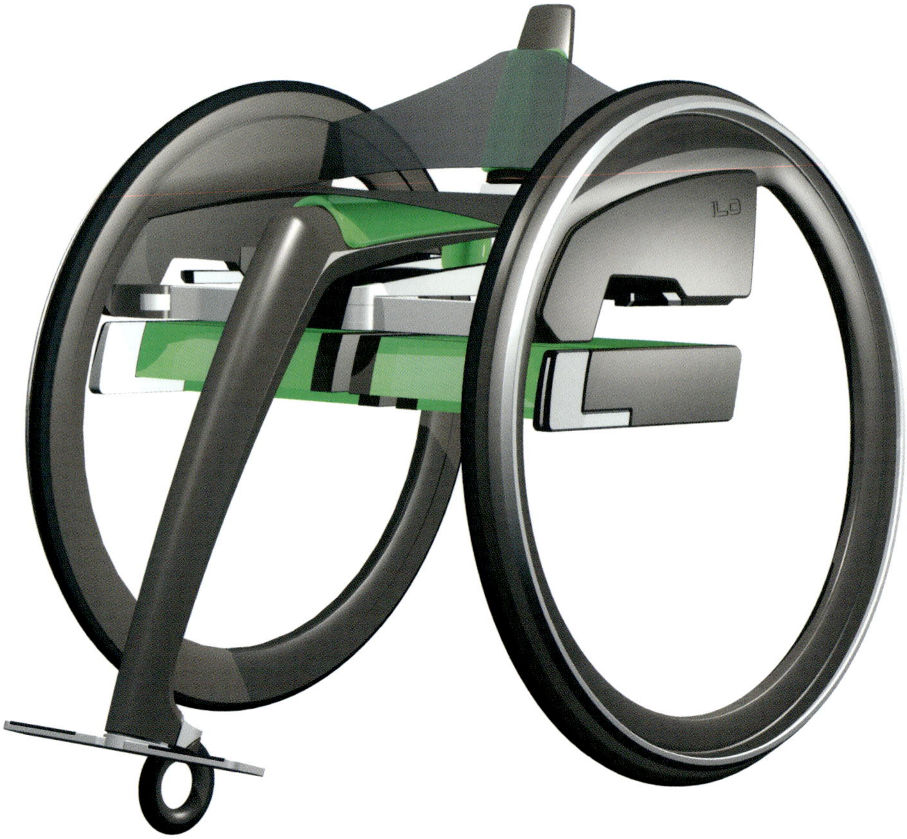

Ein Rollstuhl für aktive Behinderte, die ab der Hüfte gelähmt sind – mit dem Image- und Prestigewert eines hoch entwickelten Profigeräts. Daher der Name „Ilo" (Esperanto, d.h. Werkzeug, Instrument, Hilfsmittel). Logisch setzt er sich aus Fahrwerk, Lenkung und Sitz mit schwenkbaren Taschen zusammen.

A wheelchair for handicapped persons who are paralyzed from the hip down, but still active – with the image and prestige value of an advanced piece of professional equipment. This gives rise to its name "Ilo" (Esperanto for tool, instrument, aid). It is made up of a chassis, steering column and seat with pivoted pockets.

Jury *Judges panel*

Folgerichtig wird das Konzept aus der Analyse von sinnvollen Anforderungen in Gebrauch, Material und Ästhetik entwickelt. Durch souveräne Ausschöpfung neuer Technologien ist es gelungen, in eine wesentlich reduzierte Struktur deutlich verbesserte Funktionen zu integrieren. Zusammen mit einem technischen, dennoch die Sinne ansprechenden Formenkanon gewinnt der dreirädrige Rollstuhl den Charakter eines sportlich-funktionalen, hochwertigen Geräts, das den Nutzer in seinem Selbstwertgefühl stärkt.

The concept was developed as the logical result of an analysis of the practical demands made of use, materials and aesthetic appearance. A masterful exploitation of new technologies allows radically improved functions to be integrated in a significantly reduced structure. Together with its formal expression, which is technical yet appeals to the senses at the same time, this three-wheeler wheelchair takes on the quality of a high-grade piece of sports equipment, which boosts the user's self-confidence.

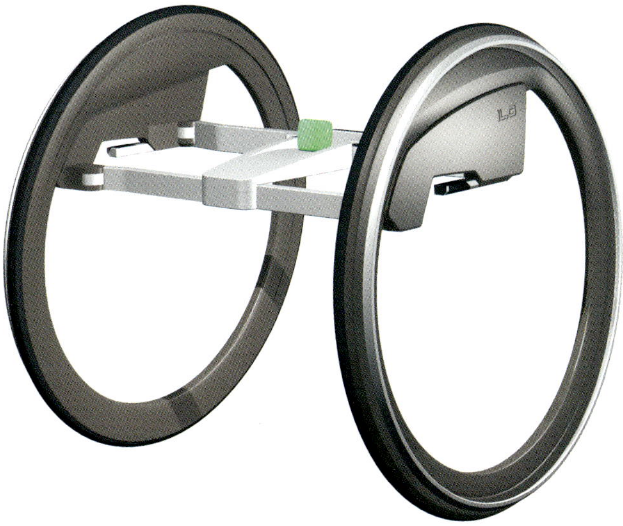

Fahrwerk: Zentraler Schemel und vier Streben aus Aluschaum, zugleich Faltmechanismus; Antriebsräder mit nabenlosen Felgen aus Karbon – leicht und stabil, kompakt zusammenfaltbar.

Chassis: central stool and four struts of aluminium foam, which also constitute the folding mechanism; drive wheels with hubless rims of carbon fibre – light and robust, compact when folded down.

Lenkung: eine Art Lenksäule aus Karbon, die von der Rückenlehne bis zum Lenkrad reicht; Fußstütze aus Aluschaum; Lenklager wie beim Skateboard – geringes Gewicht, leichtgängiges Rangieren, Steuerung durch Körperneigung.

Steering system: a kind of steering column made of carbon fibre, stretching from the back rest to the steering wheel; footrests of aluminium foam; steering bearings similar to a skateboard – low weight, easy manoeuvrability, steering by shifting weight of body.

Sitz: Sattelartige Polsterung aus TechnoGel, mit SpaceTex und GoreTex bezogen – gute Druckverteilung, verbessertes Mikroklima (Temperatur, Feuchtigkeit).

Seat: saddle-like upholstery made of TechnoGel, covered with SpaceTex and GoreTex – good distribution of pressure, improved micro-climate (temperature, moisture).

Mit einer Hand zusammenzufalten – geringer Bedarf an Stauraum.

Can be folded together with one hand, takes up little storage space.

Anerkennung (750 €) Handbuch für Polizei und Bürger
Highly commended (750 €) Manual for the police and citizens

Nikolai Affemann, Wiesbaden, nikolaiaffemann@hotmail.com, Fachhochschule Wiesbaden
Alexander Bayer, Darmstadt, alex.b.99@gmx.de, Fachhochschule Wiesbaden

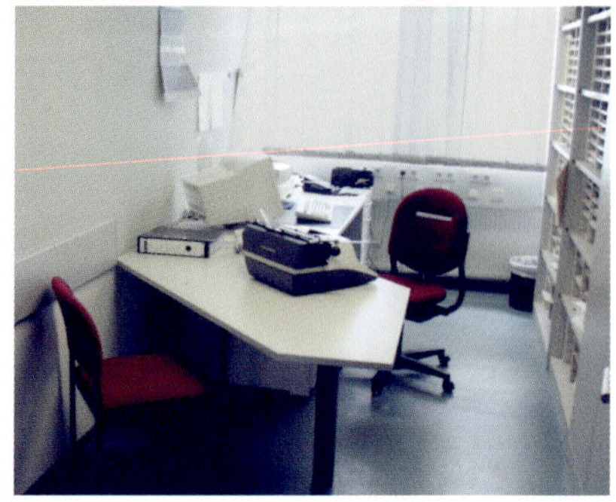

Schleuse *Security booth*

Wartebereich *Waiting room*

Vernehmungsraum *Interrogation room*

Auf der Grundlage einer in vertrauensvoller Zusammenarbeit mit den Behörden durchgeführten Analyse und Dokumentation von 15 Polizeistationen in Hessen wurde ein gestalterischer Leitfaden entwickelt. Für Um- und Neubauvorhaben gibt er Richtlinien zu Programm, Struktur, Größe und Einrichtung der für den Bürger relevanten Räume: Eingang, Schleuse, Warte-, Vernehmungs- und Opferbetreuungsraum sowie Gewahrsamszelle. Ergebnis ist ein Handbuch in gedruckter Form oder auf CD-ROM für alle am Planungsprozeß Beteiligten.

These design guidelines were developed on the basis of an analysis and documentation of 15 police stations in Hesse, conducted in cooperation with the authorities. For conversion and new construction projects, they provide a guide on type, number, structure, size and furnishings of the rooms relevant to citizens: entrance, security booth, waiting room, interrogation room and victim counselling room, and police cell. The result is a manual, either printed or on CD-ROM, for all those involved in the planning process.

Jury *Judges panel*

Die Autoren haben einen heiklen, gesellschaftlich aber wichtigen Problembereich aufgegriffen. Die Ergebnisse der Recherche sind sachlich und prägnant dargelegt. Für die als typisch erkannten Funktionsbereiche einer Polizeistation werden gestalterische Lösungen angeboten, die vom Raumprogramm bis zur Möblierung reichen und einen deutlichen Beitrag zur Bürgerfreundlichkeit der Polizei leisten. Bis auf ein multifunktionales Wandpaneel wird dabei auf eigens anzufertigende Einrichtungselemente bewußt verzichtet, damit die Hürden für eine Umsetzung möglichst niedrig sind.

The authors have tackled a tricky, yet socially important problem area. The research results are presented objectively and concisely. Design solutions are presented for those functional areas of a police station that are regarded as typical, ranging from the type and number of rooms to furnishing, and make a significant contribution to good relations between the police and citizens. Apart from a multi-functional wall panel, the guidelines consciously avoid fixtures that have to be specially produced, thus keeping the obstacles to implementation as low as possible.

Anerkennung (750 €) **Rettung aus der Luft**

Highly commended *(750 €)* ***Rescue from the air***

Manuel Aydt, Pforzheim, aydt@formverlauf.de, Hochschule für Gestaltung Technik und Wirtschaft FH Pforzheim

„RotoPro" ist ein kompakter Einmann-Helikopter in Leichtbauweise (Aluminium, Kohlefaser) für die Rettung von Verunglückten in Situationen, in denen konventionelle Hubschrauber nicht helfen können. Mit unterschiedlichen Landefüßen kann das Fluggerät auf starkem Gefälle oder zu Wasser landen, sich auch an Felswände oder Gebäudefassaden pressen. Ein Behälter nimmt die transportbedürftige Person auf. Anbauteile wie Kamera, Scheinwerfer oder Greifarm erweitern den Einsatzbereich, z.B. für Sicherheits- und Überwachungsaufgaben.

"RotoPro" is a compact, one-man, lightweight helicopter (aluminium, carbon fibre) for the rescue of victims in situations in which conventional helicopters cannot help. With various landing undercarriages the helicopter can land on steep slopes or on the water, or even press itself up against cliff faces or the walls of buildings. A container provides room for the person needing rescue. Accessories such as a camera, spotlights or a grab arm allow it to be used in other areas too, such as security or surveillance work.

Jury ***Judges panel***

Die Idee, dem Rettungswesen neue Einsatzbereiche zu erschließen, wird detail- und variantenreich ausgearbeitet. Dabei sind nicht nur die Anforderungen der Flugsicherheit und der Bedienung sorgfältig bedacht, sondern auch unterschiedliche Möglichkeiten der Bergung und des Transports. Verschiedene Ausstattungsvarianten eröffnen vielfältige Anwendungen: Wasser-, Straßen- und Bergwacht, auch kommerzielle Nutzung und Sport.

The idea of opening up new areas of operation for rescue services has been thought out in great detail and with many variants. Attention has not only been paid to the requirements of aerodynamic safety and operation, but also to the various possibilities of rescue and transport. Various accessories open up a variety of applications: coastguard, road surveillance and mountain rescue, or even commercial use and sports.

Mia Seeger Stiftung

Mia Seeger Foundation

Mia Seeger war die „Grande Dame" des Design. Vor dem Zweiten Weltkrieg wirkte sie an großen Ausstellungsprojekten des Deutschen Werkbundes mit, 1927 z.B. an der Weißenhofsiedlung in Stuttgart. Nach dem Krieg führte sie als Präsidialmitglied zwölf Jahre lang den Rat für Formgebung in Darmstadt. 1986 rief sie die nach ihr benannte Stiftung ins Leben, deren Zweck die umfassende Förderung des Design als Teil von Wirtschaft und Kultur ist. Namhafte Sponsoren aus der Wirtschaft haben sich ihren Zielen angeschlossen.
Vorsitzender des Stiftungsvorstands ist Dr. Friedrich Bullinger, Präsident des Landesgewerbeamts Baden-Württemberg, Vorsitzende des Beirats Prof. Karin Kirsch, Stuttgart.
Mit der Absicht, besonders den Nachwuchs im Design zu fördern und ihn dabei zur Auseinandersetzung mit sozialen Fragen herauszufordern, schreibt die Stiftung jährlich den Mia Seeger Preis bundesweit aus, zusätzlich das Mia Seeger Stipendium zur Förderung von Projekten.

*Mia Seeger was the "grand old lady" of design. Before the Second World War, she was involved in major exhibition projects of the Deutscher Werkbund, such as the Weissenhof complex in Stuttgart in 1927. After the war, she sat for twelve years on the managing committee of the Rat für Formgebung in Darmstadt. In 1986, she established the foundation that bears her name, whose purpose is the comprehensive promotion of design as a part of economic and cultural life. Prominent sponsors from the business world have lent their support to her aims.
The chair of the foundation's management board is Dr. Friedrich Bullinger, President of the Baden-Württemberg Office of Trade and Industry, while Prof. Karin Kirsch, Stuttgart, chairs its advisory board.
With the aim of promoting young designers in particular, and of providing them with the additional challenge of coming to terms with social problems, the Foundation invites entrants from all over Germany to compete for the annual Mia Seeger Prize. There is also a Mia Seeger Scholarship to support projects.*

Mia Seeger (1903 – 1991)

Informationen *Information*

Ausführliche Informationen zu Stiftung, Preis oder Stipendium über Internet www.mia-seeger.de oder bei der Geschäftsstelle.

For more details about the Foundation, the Prize or the Scholarship, visit www.mia-seeger.de or contact the Foundation office.

Stuttgart 2002

Anschrift *Address*

Mia Seeger Stiftung
c/o Design Center Stuttgart
Willi-Bleicher-Straße 19
70174 Stuttgart

Telefon: *Telephone:*
+49 711 123-26 84 / -26 38

Fax: *Fax:*
+49 711 123 25 77

E-Mail: *e-mail:*
design@lgabw.de

Impressum *Publishing details*

Herausgeber: *Published by:*
Mia Seeger Stiftung

Redaktion: *Editor:*
Wolfgang Berger

Grafikdesign: *Graphic design:*
avcommunication

Ausstellungsgestaltung: *Exhibition design:*
Wolfgang Frank

Plakat: *Poster:*
Uwe Lohrer

Fotos: *Photographs:*
Preisträger und Ausgezeichnete

Hersteller *Manufacturers*

Alu Meier
Kistlerhofstraße 140
D-81379 München
Tel. +49 89 7 48 90 50
Fax +49 89 74 89 05 20
info@alu-meier.de
www. alu-meier.de
Seite page 109

Artemide S.p.A.
Via Bergamo 18
I-20010 Pregnana Milanese
Tel. +39 02 93 51 81
Fax +39 02 93 59 02 54
www.artemide.com
Seite page 175

Auerhahn Bestecke GmbH
Postfach 1161
D-72206 Altensteig
Tel. +49 74 53 94 68 34
Fax +49 74 53 94 68 93
www.auerhahn-bestecke.de
Seite page 75, 99

BabyBjörn AB
Box 595
S-18236 Danderyd
Tel. +46 8 54 49 68 00
Fax +46 8 54 49 68 29
annika.sander@babybjorn.se
www.babybjorn.com
Seite page 73

Baumann + Trapp GbR
Mattenmühle 1
D-77876 Kappelrodeck
Tel. +49 78 41 29 03 00
Fax +49 78 41 29 03 05
ifkka@t-online.de
www.ideen-fuer-kinder.com
Seite page 115, 117

Bedfont LTD.
105 Laker Road,
Rochester Airport ind. Estate
Rochester, Kent ME1 3QX
Tel. +44 16 34 67 37 20
www.bedfont.com
Seite page 125

Behindertenwerkstatt Reinsdorf
gemeinnützige GmbH
Gabelsberger Straße 8
D-08141 Reinsdorf
Tel. +49 3 75 2 77 59 22
Fax +49 3 75 2 77 59 19
Seite page 29

Robert Bosch Hausgeräte GmbH
Hochstraße 17
D-81669 München
Tel. +49 89 45 90 01
Fax +49 89 45 90 23 47
Seite page 49, 57

Brabantia
Brabantialaan 3
NL-5583 BP Waalre
Tel. +31 40 2 28 22 58
Fax +31 40 2 28 23 82
c.jorissen@brabantia.com
www.brabantia.com
Seite page 101

Bree Collection GmbH & Co. KG
Gerberstraße 3
D-30916 Isernhagen
Tel. +49 51 36 8 97 60
Fax +49 51 36 8 97 62 29
n.barth@bree.de
www.bree.de
Seite page 123

Brüggli Produktion & Dienstleistung
Hofstraße 5
CH-8590 Romanshorn
Tel. +41 71 4 66 94 94
Fax +41 71 4 66 94 95
daniel.brunner@brueggli.ch
www.leggero.com
Seite page 189

Casco Schutzhelme GmbH
Gewerbering Süd 11
D-01900 Bretnig-Hauswalde
Tel. +49 3 59 55 83 90
Fax +49 3 59 55 8 39 99
casco-schutzhelme@t-online.de
www.casco-schutzhelme.de
Seite page 105

Ceka Büromöbelwerke
Erich-Krause-Straße 1
D-36304 Alsfeld
Tel. +49 66 31 18 62 62
www.ceka.de
Seite page 139

Chroma Cutlery Inc.
184 Oakland Ave. Suite B
Spartanburg SC 29302, USA
Tel. +1 8 88 2 50 88 09
Fax +1 8 01 2 18 89 89
www.happy-cooking.com
Seite page 81

Dalli-Klick
Josef Sekler
Hasenbergstraße 24
D-73479 Ellwangen-Pfahlheim
Tel. +49 79 65 80 06 32
Fax +49 12 12 5 10 51 79 38
wirsekler@t-online.de
www.wir-einrichtungen.de
Seite page 141

Dataplus GmbH, Kratzert & Schrem
Hesselbühl 5
D-88630 Pfullendorf
Tel. +49 75 52 2 60 90
Fax +49 75 52 26 09 10
info@dataplus.de
www.dataplus.de
Seite page 165, 167

Do!
Falkoner Allé 44, 4th
DK-2000 Frederiksberg
Tel. +45 40 19 23 35
designs@do-co.dk
www.do-co.dk
Seite page 161

Dornbracht GmbH & Co. KG
Kobbingsmühle 6
D-58640 Iserlohn
Tel. +49 23 71 43 30
Fax +49 23 71 43 32 32
Seite page 65

Duravit AG
Werderstraße 36
D-78132 Hornberg
Tel. +49 78 33 7 00
Fax +49 78 33 96 09 02
duravit@duravit.de
www.duravit.de
Seite page 63

Eberlein GmbH
Ansbacher Straße 126
D-90449 Nürnberg
Seite page 145

Engelbreit & Sohn
Marco Engelbreit
Heinrich-Diehl-Straße 25
D-90552 Röthenbach
Seite page 145

Eva Denmark A/S
Hoejnaesvej 59
DK-2610 Roedovre
Tel. +45 36 73 20 60
Fax +45 36 70 74 11
mail@evadenmark.com
www.evadenmark.com
Seite page 93, 95

Fiskars Consumer Oy Ab
FIN-10330 Billnäs
Tel. +3 58 19 27 77 21
Fax +3 58 19 23 09 86
www.fiskars.fin
Seite page 181

Formfarm
Neuer Berg 24
D-70327 Stuttgart
Tel. +49 7 11 3 04 00 70
Fax +49 7 11 3 04 00 72
info@formfarm.de
www.formfarm.de
Seite page 143

Tobias Grau GmbH
Siemensstraße 35B
D-25462 Rellingen
Tel. +49 41 01 37 00
Fax +49 41 01 3 70 10 00
www.tobias-grau.com
Seite page 171, 173

Grundig AG
Beuthener Straße 41
D-90471 Nürnberg
Tel. +49 9 11 7 03 72 53
Fax +49 9 11 7 03 75 89
juergen.schoenborn@grundig.com
www.grundig.com
Seite page 201

Hansgrohe AG
Auestraße 5–9
D-77761 Schiltach
Tel. +49 78 36 51 14 93
Fax +49 78 36 5 14 14 93
info@hansgrohe.com
www.hansgrohe.com
Seite page 67

Matthias Hohner AG
Hohnerstraße 8
D-78647 Trossingen
Tel. +49 74 25 2 00
Fax +49 74 25 2 03 71
Seite page 113

Imperial-Werke oHG
Installstraße 10–18
D-32257 Bünde
Tel. +49 52 23 48 12 23
Fax +49 52 23 48 13 36
www.imperial.de
Seite page 47

Indigo Pure Snowboards
Thorsten Schwabe & Gregor Baer
Blutenburgstraße 41a
D-80636 München
Tel. +49 89 18 95 45 50
Fax +49 89 18 95 45 49
kontakt@indigosnow.de
www.indigosnow.de
Seite page 107

Invair Technologies AG
Wollgrasweg 49
D-70599 Stuttgart
Tel. +49 1 60 7 14 14 83
Fax +49 7 11 4 51 01 75 19
s.schmidt@invair.ch
www.invair.com
Seite page 203

Jahn Design Fact
Schlesische Straße 27
D-10997 Berlin
Tel. +49 30 5 30 00 60
Fax +49 30 53 00 06 66
hj@jahn-design.com
www.jahn-design.com
Seite page 33

Möbelbau Kaether & Weise GmbH
Dammstraße 43
D-31195 Lamspringe
Tel. +49 51 83 6 77
Fax +49 51 83 27 66
kaetherundweise@t-online.de
www.kaetherundweise.de
Seite page 35

Klöber GmbH
Rauensteinstraße 18
D-88662 Überlingen
Tel. +49 75 51 83 81 46
Fax +49 75 51 83 81 42
info@kloeber.com
www.kloeber.com
Seite page 133

Walter Knoll AG & Co. KG
Bahnhofstraße 25
D-71032 Herrenberg
Tel. +49 70 32 20 80
Fax +49 70 32 20 82 54
info@walterknoll.de
www.walterknoll.de
Seite page 41

C. Josef Lamy GmbH
Grenzhöfer Weg 32
D-69123 Heidelberg
Tel. +49 62 21 84 31 04
Fax +49 62 21 84 31 21
wweber@lamy.de
www.lamy.de
Seite page 159

Lather Kommunikation
Fronhäuser Straße 19
D-35102 Lohra-Damm
Tel. +49 64 26 9 24 10
Fax +49 64 26 92 41 50
info@lather.de
www.lather.de
Seite page 163

Leitner GmbH
Düsseldorfer Straße 14
D-71332 Waiblingen
Tel. +49 71 51 1 70 60
Fax +49 71 51 17 06 76
system@leitner.de
www.leitner.de
Seite page 137

Lemonfish
Alexandra Dittrich +
Bettina Weitershagen GbR
Lindenfirststraße 14/1
D-73525 Schwäbisch Gmünd
Tel. +49 71 71 40 51 19
Fax +49 71 71 40 51 24
info@lemonfish.de
www.lemonfish.de
Seite page 121

LRG. Licht + Raum AG
Hinterer Schermen 44
CH-3063 Ittingen
Tel. +41 31 9 24 77 88
Fax +41 31 9 24 77 80
LRG@bluewin.ch
Seite page 177

MAM Babyartikel GmbH
Lorenz-Mandl-Gasse 50
A-1160 Wien
Tel. +43 1 49 14 10
Fax +43 1 49 14
office@mambaby.com
www. mambaby.com
Seite page 69

Merten GmbH & Co. KG
Fritz-Kotz-Straße 8
D-51674 Wiehl
Tel. +49 22 61 7 02 01
Fax +49 22 61 70 22 84
info@merten.de
www.merten.de
Seite page 197

Nils Holger Moormann
Möbel-Produktions- und Handels GmbH
Festhalle
D-83229 Aschau i. Chiemgau
Tel. +49 80 52 9 04 50
Fax +49 80 52 90 45 45
info@moormann.de
www.moormann.de
Seite page 43

Naef Spiele AG
Grendelweg 5
CH-4314 Zeiningen
Tel. +41 61 8 51 18 44
Fax +41 61 8 51 18 41
www.naefspiele.ch
Seite page 119

Neff GmbH
Ruiterstraße 8
D-75015 Bretten
Tel. +49 72 52 97 64 04
Fax +49 72 52 97 65 80
Seite page 53, 55

Performa Möbel und Design GmbH
Marbacher Straße 54
D-74385 Pleidelsheim
Tel. +49 71 44 89 88 69
Fax +49 71 44 89 88 76
u.walz@performa.de
www.performa.de
Seite page 39, 135

Dachziegelwerke
Pfleiderer GmbH & Co. KG
Marbacher Straße 4
D-71364 Winnenden
Tel. +49 71 95 69 60
Fax +49 71 95 69 62 25
ranssweiler@pfleiderer-dach.de
www.pfleiderer-dach.de
Seite page 199

C. Hugo Pott GmbH & Co. KG
Ritterstraße 28
D-42659 Solingen
Tel. +49 2 12 4 30 56
Fax +49 2 12 4 24 25
pott@pott-bestecke.de
www.pott-bestecke.de
Seite page 83

Protrade Europe
Pforzheimer Straße 63
D-75449 Wurmberg
Tel. +49 70 44 94 51 20
Fax +49 70 44 94 51 30
protrade@protrade.de
www.protrade.de
Seite page 185

Pyramid Computer Systeme GmbH
Bötzinger Straße 60
D-79111 Freiburg
Tel. +49 7 61 4 51 47 11
Fax +49 7 61 45 14 70
fh@pyramid.de
www.pyramid.de
Seite page 205

Reisenthel Accessoires
Benzstraße 3
D-82178 Puchheim
Tel. +49 89 8 00 64 40
Fax +49 89 80 60 35
Seite page 147, 151

Gebr. Richartz + Söhne GmbH
Merscheider Straße 94
D-42699 Solingen
Tel. +49 2 12 23 23 10
Fax +49 2 12 2 32 31 99
stefanrichartz@richartz.com
www.richartz.com
Seite page 85

Rieber GmbH & Co. KG
Hoffmannstraße 44
D-72770 Reutlingen
Tel. +49 71 21 51 86 40
Fax +49 71 21 51 86 49
info@rieber.de
www.rieber.de
Seite page 187

Ritterwerk GmbH, Hausgeräte
Industriestraße 13
D-82194 Gröbenzell
Tel. +49 81 42 5 10 41
Fax +49 81 42 5 82 23
marketing@ritterwerk.de
www.ritterwerk.de
Seite page 97

Rosenthal AG
Wittelsbacherstraße 43
D-95100 Selb
Tel. +49 92 87 7 23 44
Fax +49 92 87 7 22 71
silke-troeger@rosenthal.de
www.rosenthal.de
Seite page 89, 91

Rupp + Hubrach Optik GmbH
Von-Ketteler-Straße 1
D-96050 Bamberg
Tel. +49 9 51 1 86 13 31
Fax +49 9 51 1 86 13 69
b.stoermann@rh-brillenglas.de
www.rh-brillenglas.de
Seite page 207

Sammet Moebelkonzepte
Ellenbergstraße 19
D-71729 Erdmannhausen
Tel. +49 71 48 16 24 77
frank.sammet@bueroform.net
Seite page 179

Alfred Scholze GmbH
Gürtlerstraße 6
D-87600 Kaufbeuren, Neugablonz
Tel. +49 83 41 9 77 00
Fax +49 83 41 6 29 32
Seite page 87

Schwan-Stabilo
Schwanhäußer GmbH & Co.
Schwanweg 1
D-90562 Heroldsberg
Tel. +49 9 11 5 67 13 65
Fax +49 9 11 5 67 13 11
www.stabilo.com
Seite page 153, 155

Sedus Stoll AG
Brückenstraße 15
D-79761 Waldshut
Tel. +49 77 51 8 42 47
Fax +49 77 51 8 42 45
MathiasSeiler@sedus.de
www.sedus.de
Seite page 131

Siemens Electrogeräte GmbH
Hochstraße 17
D-81669 München
Tel. +49 89 45 90 23 87
Fax +49 89 45 90 29 58
www.siemens-hausgeraete.de
Seite page 51, 59, 61

Staedtler Mars GmbH & Co
Moosäckerstraße 3
D-90427 Nürnberg
Tel. +49 9 11 9 36 50
Fax +49 9 11 9 36 54 00
info@staedtler.de
www.staedtler.de
Seite page 157

Tojo Möbel GmbH
Am Ziegelgraben 26
D-73614 Schorndorf
Tel. +49 71 81 40 14 00
Fax +49 71 81 40 14 10
info@tojo.de
www.tojo.de
Seite page 37

Tonon & C. S.p.A.
Via Diaz 22
I-33044 Manzano
Tel. +39 04 32 74 07 40
Fax +39 04 32 74 07 70
Seite page 129

Uwe GmbH
Buchstraße 82
D-73525 Schwäbisch Gmünd
Tel. +49 71 71 10 32 16
info@uwe-germany.de
Seite page 111

Veigel GmbH + Co. KG
Lindenstraße 9–11
D-74653 Künzelsau
Tel. +49 79 40 91 30 17
Fax +49 79 40 91 30 60
Hinrich.Swyter@veigel-automotive.de
www.veigel-automotive.de
Seite page 191

Viessmann Werke
D-35107 Allendorf
Tel. +49 64 52 70 23 31
Fax +49 64 52 70 53 31
sgl@viessmann.com
www.viessmann.com
Seite page 195

Walz GmbH & Co
Im Lehrer Feld 6
D-89081 Ulm
Tel. +49 7 31 96 27 70
Fax +49 7 31 9 62 77 74
info@walz-ulm.de
www.walz-ulm.de
Seite page 183

M. H. Wilkens & Söhne GmbH
An der Silberpräge
D-28305 Bremen
Tel. +49 4 21 4 10 32 73
Fax +49 4 21 4 10 31 30
Seite page 79

WMF Württembergische Metallwaren-
fabrik AG
Eberhardstraße
D-73312 Geislingen/Steige
Tel. +49 73 31 2 51
Fax +49 73 31 4 53 87
t.schuetz@wmf.de
www.wmf.de
Seite page 77

Vertriebe *Distributors*

Artemide GmbH
Hans-Böckler-Straße 2
D-58730 Fröndenberg
Tel. +49 23 73 97 50
Fax +49 23 73 97 52 24
ml@artemide.de
www.artemide.com
Seite page 175

Astra Licht- und Deckensysteme Vertriebs GmbH
Talstraße 19
D-70825 Korntal-Münchingen
Tel. +49 7 11 8 39 50 00
Fax +49 7 11 83 95 00 22
Seite page 177

Auerhahn Bestecke GmbH
Postfach 1161
D-72206 Altensteig
Tel. +49 74 53 94 68 34
Fax +49 74 53 94 68 93
www.auerhahn-bestecke.de
Seite page 75, 99

B/S/H/ München CNLK - Z
Hochstraße 17
D-81669 München
Seite page 53, 55

Baumann + Trapp GbR
Mattenmühle 1
D-77876 Kappelrodeck
Tel. +49 78 41 29 03 00
Fax +49 78 41 29 03 05
ifkka@t-online.de
www.ideen-fuer-kinder.com
Seite page 115, 117

Behindertenwerkstatt Reinsdorf gemeinnützige GmbH
Gabelsberger Straße 8
D-08141 Reinsdorf
Tel. +49 3 75 2 77 59 22
Fax +49 3 75 2 77 59 19
Seite page 29

Berendsohn AG
Alte Sülldorfer Landstraße 400
D-22559 Hamburg
Tel. +49 40 81 90 90
Fax +49 40 81 90 92 96
aoffen@berendsohn.com
www.berendsohn.com
Seite page 87

Birkhold KG
Vogelsangstraße 16a
D-70176 Stuttgart
Tel. +49 7 11 66 98 50
Fax +49 7 11 6 69 85 90
Seite page 203

Robert Bosch Hausgeräte GmbH
Hochstraße 17
D-81669 München
Tel. +49 89 45 90 01
Fax +49 89 45 90 23 47
Seite page 49, 57

Ceka Büromöbelwerke
Erich-Krause-Straße 1
D-36304 Alsfeld
Tel. +49 66 31 18 60
www.ceka.de
Seite page 139

CRR Christian Romanowski Rockstedt e.K.
Haupststraße 16
D-27404 Rockstedt
Tel. +49 18 03 59 59 59
Fax +49 18 03 74 74 74
beste@kuechenmesser.de
www.kochmesser.de
Seite page 81

Czechau
Breite Straße 68
D-32657 Lemgo
Tel. +49 52 61 39 39
Fax +49 52 61 1 55 15
Seite page 161

Dalli-Klick
Josef Sekler
Hasenbergstraße 24
D-73479 Ellwangen-Pfahlheim
Tel. +49 79 65 80 06 32
Fax +49 12 12 5 10 51 79 38
wirsekler@t-online.de
www.wir-einrichtungen.de
Seite page 141

Dataplus GmbH, Kratzert & Schrem
Hesselbühl 5
D-88630 Pfullendorf
Tel. +49 75 52 2 60 90
Fax +49 75 52 26 09 10
info@dataplus.de
www.dataplus.de
Seite page 165, 167

3L Modulbausystem
Jürgen Steinhauser
Steinstraße 21
D-90419 Nürnberg
Tel. +49 9 11 77 60 70
Fax +49 9 11 77 60 80
js@3LM.de
www.loftschau.de
Seite page 145

Duravit AG
Werderstraße 36
D-78132 Hornberg
Tel. +49 78 33 7 00
Fax +49 78 33 96 09 02
duravit@duravit.de
www.duravit.de
Seite page 63

Eva Deutschland
Pinneberger Chaussee 99
D-25436 Moorrege
Tel. +49 41 22 8 18 62
Fax +49 41 22 8 92 22
Seite page 93, 95

Fiskars Deutschland GmbH
Weststraße 42
D-40721 Hilden
Tel. +49 21 03 5 89 00
Fax +49 21 03 58 90 45
Seite page 181

Formfarm
Neuer Berg 24
D-70327 Stuttgart
Tel. +49 7 11 3 04 00 70
Fax +49 7 11 3 04 00 72
info@formfarm.de
www.formfarm.de
Seite page 143

Tobias Grau GmbH
Siemensstraße 35B
D-25462 Rellingen
Tel. +49 41 01 37 00
Fax +49 41 01 3 70 10 00
www.tobias-grau.com
Seite page 171, 173

Grundig AG
Beuthener Straße 41
D-90471 Nürnberg
Tel. +49 9 11 7 03 72 53
Fax +49 9 11 7 03 75 89
juergen.schoenborn@grundig.com
www.grundig.com
Seite page 201

Hansgrohe AG
Auestraße 5–9
D-77761 Schiltach
Tel. +49 78 36 51 14 93
Fax +49 78 36 5 14 14 93
info@hansgrohe.com
www.hansgrohe.com
Seite page 67

Indigo Pure Snowboards
Thorsten Schwabe & Gregor Baer
Blutenburgstraße 41a
D-80636 München
Tel. +49 89 18 95 45 50
Fax +49 89 18 95 45 49
kontakt@indigosnow.de
www.indigosnow.de
Seite page 107

Jahn Design Fact
Schlesische Straße 27
D-10997 Berlin
Tel. +49 30 5 30 00 60
Fax +49 30 53 00 06 66
hj@jahn-design.com
www.jahn-design.com
Seite page 33

Möbelbau Kaether & Weise GmbH
Dammstraße 43
D-31195 Lamspringe
Tel. +49 51 83 6 77
Fax +49 51 83 27 66
kaetherundweise@t-online.de
www.kaetherundweise.de
Seite page 35

Walter Knoll AG & Co. KG
Bahnhofstraße 25
D-71032 Herrenberg
Tel. +49 70 32 20 80
Fax +49 70 32 20 82 54
info@walterknoll.de
www.walterknoll.de
Seite page 41

C. Josef Lamy GmbH
Grenzhöfer Weg 32
D-69123 Heidelberg
Tel. +49 62 21 84 31 04
Fax +49 62 21 84 31 21
wweber@lamy.de
www.lamy.de
Seite page 159

Lather Kommunikation
Fronhäuser Straße 19
D-35102 Lohra-Damm
Tel. +49 64 26 9 24 10
Fax +49 64 26 92 41 50
info@lather.de
www.lather.de
Seite page 163

Leitner GmbH
Düsseldorfer Straße 14
D-71332 Waiblingen
Tel. +49 71 51 1 70 60
Fax +49 71 51 17 06 76
system@leitner.de
www.leitner.de
Seite page 137

Lemonfish
Alexandra Dittrich +
Bettina Weitershagen GbR
Lindenfirststraße 14/1
D-73525 Schwäbisch Gmünd
Tel. +49 71 71 40 51 19
Fax +49 71 71 40 51 24
info@lemonfish.de
www.lemonfish.de
Seite page 121

MAM Babyartikel GmbH
Rudolf-Diesel-Straße 6
D-27383 Scheessel
Tel. +49 42 63 9 31 70
Fax +49 42 63 93 17 20
Seite page 69

Merten GmbH & Co. KG
Fritz-Kotz-Straße 8
D-51674 Wiehl
Tel. +49 22 61 7 02 01
Fax +49 22 61 70 22 84
info@merten.de
www.merten.de
Seite page 197

Metris Planungsgesellschaft GbR
Rhönring 121
64289 Darmstadt
Tel. +49 61 51 99 77 17
Fax +49 61 51 99 78 35
www.metris-office.de
Seite page 31

Musik Meyer GmbH
Industriestraße 20
D-35041 Marburg
Tel. +49 64 21 98 90
Fax +49 64 21 98 93 59
Seite page 113

Nils Holger Moormann
Möbel-Produktions- und Handels GmbH
Festhalle
D-83229 Aschau i. Chiemgau
Tel. +49 80 52 9 04 50
Fax +49 80 52 90 45 45
info@moormann.de
www.moormann.de
Seite page 43

Naef Spiele AG
Grendelweg 5
CH-4314 Zeiningen
Tel. +41 61 8 51 18 44
Fax +41 61 8 51 18 41
www.naefspiele.ch
Seite page 119

Neomed GmbH
Danzigerstraße 93
D-41352 Korschenbroich
Tel. +49 21 61 64 42 19
Fax +49 21 61 64 47 46
Seite page 125

Performa Möbel und Design GmbH
Marbacher Straße 54
D-74385 Pleidelsheim
Tel. +49 71 44 89 88 69
Fax +49 71 44 89 88 76
u.walz@performa.de
www.performa.de
Seite page 39, 135

Dachziegelwerke
Pfleiderer GmbH & Co. KG
Marbacher Straße 4
D-71364 Winnenden
Tel. +49 71 95 69 60
Fax +49 71 95 69 62 25
ranssweiler@pfleiderer-dach.de
www.pfleiderer-dach.de
Seite page 199

C. Hugo Pott GmbH & Co. KG
Ritterstraße 28
D-42659 Solingen
Tel. +49 2 12 4 30 56
Fax +49 2 12 4 24 25
pott@pott-bestecke.de
www.pott-bestecke.de
Seite page 83

Protrade Europe
Pforzheimer Straße 63
D-75449 Wurmberg
Tel. +49 70 44 94 51 20
Fax +49 70 44 94 51 30
protrade@protrade.de
www.protrade.de
Seite page 185

Pyramid Computer Systeme GmbH
Bötzinger Straße 60
D-79111 Freiburg
Tel. +49 7 61 4 51 47 11
Fax +49 7 61 45 14 70
fh@pyramid.de
www.pyramid.de
Seite page 205

Reisenthel Accessoires
Benzstraße 3
D-82178 Puchheim
Tel. +49 89 8 00 64 40
Fax +49 89 80 60 35
Seite page 147, 151

Gebr. Richartz + Söhne GmbH
Merscheider Straße 94
D-42699 Solingen
Tel. +49 2 12 23 23 10
Fax +49 2 12 2 32 31 99
stefanrichartz@richartz.com
www.richartz.com
Seite page 85

Rieber GmbH & Co. KG
Hoffmannstraße 44
D-72770 Reutlingen
Tel. +49 71 21 51 80
Fax +49 71 21 51 83 02
info@rieber.de
www.rieber.de
Seite page 187

Ritterwerk GmbH, Hausgeräte
Industriestraße 13
D-82194 Gröbenzell
Tel. +49 81 42 5 10 41
Fax +49 81 42 5 82 23
marketing@ritterwerk.de
www.ritterwerk.de
Seite page 97

Rupp + Hubrach Optik GmbH
Von-Ketteler-Straße 1
D-96050 Bamberg
Tel. +49 9 51 1 86 13 31
Fax +49 9 51 1 86 13 69
b.stoermann@rh-brillenglas.de
www.rh-brillenglas.de
Seite page 207

Sales Plus BabyBjörn Deutschland
Wetterkreuz 5
D-91058 Erlangen
Tel. +49 1 80 55 16 00
Seite page 73

Sammet Moebelkonzepte
Ellenbergstraße 19
D-71729 Erdmannhausen
Tel. +49 71 48 16 24 77
frank.sammet@bueroform.net
Seite page 179

Schwan-Stabilo Marketing
Deutschland GmbH & Co.
Schwanweg 1
D-90562 Heroldsberg
Tel. +49 9 11 5 67 32 66
Fax +49 9 11 5 67 44 44
www.stabilo.com
Seite page 153, 155

Sedus Stoll AG
Brückenstraße 15
D-79761 Waldshut
Tel. +49 77 51 8 42 47
Fax +49 77 51 8 42 45
MathiasSeiler@sedus.de
www.sedus.de
Seite page 131

Siemens Electrogeräte GmbH
Hochstraße 17
D-81669 München
Tel. +49 89 45 90 01
Fax +49 89 45 90 23 47
www.siemens-hausgeraete.de
Seite *page 51, 59, 61*

Staedtler Deutschland Vertriebs-GmbH
Moosäckerstraße 3
D-90427 Nürnberg
Tel. +49 9 11 9 36 52 32
Fax +49 9 11 9 36 52 30
Seite *page 157*

Tojo Möbel GmbH
Am Ziegelgraben 26
D-73614 Schorndorf
Tel. +49 71 81 40 14 00
Fax +49 71 81 40 14 10
info@tojo.de
www.tojo.de
Seite *page 37*

Uwe GmbH
Buchstraße 82
D-73525 Schwäbisch Gmünd
Tel. +49 71 71 10 32 16
info@uwe-germany.de
Seite *page 111*

Veigel GmbH + Co. KG
Lindenstraße 9–11
D-74653 Künzelsau
Tel. +49 79 40 91 30 17
Fax +49 79 40 91 30 60
Hinrich.Swyter@veigel-automotive.de
www.veigel-automotive.de
Seite *page 191*

Viessmann Werke
D-35107 Allendorf
Tel. +49 64 52 70 23 31
Fax +49 64 52 70 53 31
sgl@viessmann.com
www.viessmann.com
Seite *page 195*

Waldhausen GmbH & Co. KG
Von-Hünefeld-Straße 53
D-50829 Köln-Ossendorf
Tel. +49 2 21 58 80 10
Fax +49 2 21 588 01 25
Seite *page 105*

Walz GmbH & Co
Im Lehrer Feld 6
D-89081 Ulm
Tel. +49 7 31 96 27 70
Fax +49 7 31 9 62 77 74
info@walz-ulm.de
www.walz-ulm.de
Seite *page 183*

M. H. Wilkens & Söhne GmbH
An der Silberpräge
D-28305 Bremen
Tel. +49 4 21 4 10 32 73
Fax +49 4 21 4 10 31 30
Seite *page 79*

WK Wohnen GmbH
Im Gefierth 9a
D-63303 Dreieich
Tel. +49 61 03 39 10
www.designo.de
Seite *page 129*

WMF Württembergische Metallwaren-fabrik AG
Eberhardstraße
D-73312 Geislingen/Steige
Tel. +49 73 31 2 51
Fax +49 73 31 4 53 87
t.schuetz@wmf.de
www.wmf.de
Seite *page 77*

Designer *Designers*

Nikolai Affemann
Philippsbergstraße 36
D-65195 Wiesbaden
Tel. +49 6 11 5 90 06 84
nikolaiaffemann@hotmail.com
Seite page 216

a/2
Thomas Albrecht
Kattenstraße 8
D-34119 Kassel
Tel. +49 5 61 7 66 76 18
Fax +49 5 61 7 66 76 49
Seite page 135

Manuel Aydt
Westliche Karl-Friedrich-Straße 147
D-75172 Pforzheim
Tel. +49 72 31 56 57 28
Mobil +49 1 63 5 52 14 51
aydt@formverlauf.de
Seite page 217

B-Plan
Rathlefstraße 7
D-31582 Nienburg/Weser
Tel. +49 50 21 92 52 28
Seite page 139

Gregor Baer, Thorsten Schwabe
Indigo Pure Snowboards
Blutenburgstraße 41a
D-80636 München
Tel. +49 89 18 95 45 50
Fax +49 89 18 95 45 49
kontakt@indigosnow.de
www.indigosnow.de
Seite page 107

Martin Ballendat
Maximilianstraße 4
D-84359 Simbach am Inn
Tel. +49 85 71 60 56 60
Fax +49 85 71 60 56 66
Seite page 129

Dieter Baumann, Wilfried Trapp
Baumann + Trapp GbR
Mattenmühle 1
D-77876 Kappelrodeck
Tel. +49 78 41 29 03 00
Fax +49 78 41 29 03 05
ifkka@t-online.de
www.ideen-fuer-kinder.com
Seite page 115, 117

Peter Bäurle
Wiesentalstraße 43/1
D-73312 Geislingen/Eybach
Tel. +49 73 31 6 44 30
Seite page 75

Alexander Bayer
Kirchstraße 32
D-64367 Mühltal
Tel. +49 61 51 14 63 54
alex.b.99@gmx.de
Seite page 216

Christoph Becke
Siemens Electrogeräte GmbH
Hochstraße 17
D-81669 München
Tel. +49 89 45 90 26 35
Fax +49 89 45 90 29 58
gerd.wilsdorf@bshg.com
www.siemens-hausgeraete.de
Seite page 59

I. D. Pool; Ernst Beranek
Singerstraße 6
A-1010 Wien
Tel. +43 1 5 12 58 41
Fax +43 1 5 12 58 41 75
Seite page 69

Bluequest Design Group
früher:
Topel und Pauser Industrial Design
Lichtenbergstraße 68
D-64289 Darmstadt
Tel. +49 61 51 9 67 28 70
Fax +49 61 51 9 67 28 28
info @bluequest.de
www.bluequest.de
Seite page 113

Bosch Werksdesign
Robert-Bosch-Straße
D-89537 Giengen
Tel. +49 73 22 92 25 52
Fax +49 73 22 92 25 12
roland.vetter@bshg.com
Seite page 49

Brabantia
Brabantialaan 3
NL-5583 BP Waalre
Tel. +31 40 2 28 22 58
Fax +31 40 2 28 23 82
c.jorissen@brabantia.com
www.brabantia.com
Seite page 101

Benjamin Gugelberger, Ulf Braun
Neuer Berg 24
D-70327 Stuttgart
Tel. +49 7 11 3 04 00 70
Fax +49 7 11 3 04 00 72
Seite page 143

Daniela Breuer, Christian Reichert
Bree Collection GmbH & Co. KG
Gerberstraße 3
D-30916 Isernhagen
Tel. +49 51 36 8 97 60
Fax +49 51 36 8 97 62 29
n.barth@bree.de
www.bree.de
Seite page 123

Werksdesign
Casco Schutzhelme GmbH
Gewerbering Süd 11
D-01900 Bretnig-Hauswalde
Tel. +49 3 59 55 83 90
Fax +49 3 59 55 8 39 99
casco-schutzhelme@t-online.de
www.casco-schutzhelme.de
Seite page 105

Ceka Werksdesign
Ceka Büromöbelwerke
Erich-Krause-Straße 1
D-36304 Alsfeld
Tel. +49 66 31 18 62 62
www.ceka.de
Seite page 139

Chrétien + Apothéloz
Lägernstraße 27
CH-8037 Zürich
Tel. +41 1 3 61 51 47
Fax +41 1 3 61 51 97
welcome@industrialdesigners.ch
www.industrialdesigners.ch
Seite page 189

Claudia Christl
M. H. Wilkens & Söhne GmbH
An der Silberpräge
D-28305 Bremen
Tel. +49 4 21 4 10 32 45
Fax +49 4 21 4 10 31 30
Seite page 79

Waltraud Claar
Untere Bergstraße 3
D-71334 Waiblingen
Tel. +49 71 51 3 45 54
Fax +49 71 51 50 86 26
Seite page 187

Franco Clivio
Rietstraße 11
CH-8703 Erlenbach
Tel. +41 19 12 06 60
Fax +41 19 12 06 62
Seite page 159

Natalie Crasset
c/o Artemide S.p.A.
Via Bergamo 18
I-20010 Pregnana Milanese
Seite page 175

Karl Dittert
Eichenweg 32
D-73525 Schwäbisch Gmünd
Tel. +49 71 71 6 22 63
Fax +49 71 71 6 12 60
Seite page 97

Lemonfish
Alexandra Dittrich +
Bettina Weitershagen GbR
Lindenfirststraße 14/1
D-73525 Schwäbisch Gmünd
Tel. +49 71 71 40 51 19
Fax +49 71 71 40 51 24
info@lemonfish.de
www.lemonfish.de
Seite page 121

Werner Engels
Grundig AG
Beuthener Straße 41
D-90471 Nürnberg
Tel. +49 9 11 70 30
Fax +49 9 11 7 03 75 89
www.grundig.com
Seite page 201

Ergonomi Design Gruppen
Håkan Bergkvist
Box 14021
S-16141 Bromma
Tel./Fax +46 8 26 22 41
Seite page 73

Factor Product
Boris Simon, Frank Thiele
Comeniusstraße 1
D-81667 München
Tel. +49 89 48 92 78 10
Fax +49 89 48 92 78 11
Seite page 153

Formgestaltung Schnell-Waltenberger
Sponheimstraße 22
D-75177 Pforzheim
Tel./Fax +49 72 31 3 33 64
Seite page 205

Thorsten Franck
Lautensackstraße 21
D-80687 München
Tel. +49 89 57 86 91 94
Fax +49 89 57 86 91 95
Seite page 35

Lutz Gathmann
Nagelsweg 41
D-40474 Düsseldorf
Tel. +49 2 11 45 29 19
Fax +49 2 11 45 29 06
info@lutz-gathmann.de
www.lutz-gathmann.de
Seite page 185, 207

Johannes Geyer, Gerd E. Wilsdorf
Siemens Electrogeräte GmbH
Hochstraße 17
D-81669 München
Tel. +49 89 45 90 21 59
Fax +49 89 45 90 29 58
gerd.wilsdorf@bshg.com
www.siemens-hausgeraete.de
Seite page 61

Tobias Grau
Tobias Grau GmbH
Siemensstraße 35B
D-25462 Rellingen
Tel. +49 41 01 37 00
Fax +49 41 01 3 70 10 00
www.tobias-grau.com
Seite page 171, 173

Nicholas Grimshaw & Partners
Industrial Design
1 Conway Street, Fitzroy Square
London W1P 6LR
Tel. +44 20 72 91 41 41
Fax +44 20 72 91 41 94
Seite page 197

Ralf Grobleben
Ruiterstraße 8
D-75015 Bretten
Tel. +49 72 52 97 61 49
Fax +49 72 52 97 65 30
Seite page 55

Joachim Grützke
Ruiterstraße 8
D-75015 Bretten
Tel. +49 72 52 97 63 14
Fax +49 72 52 97 65 30
Seite page 53

Benjamin Gugelberger, Ulf Braun
Neuer Berg 24
D-70327 Stuttgart
Tel. +49 7 11 3 04 00 70
Fax +49 7 11 3 04 00 72
Seite page 143

Rolf Heide
Drosselweg 5
D-22926 Ahrensburg
Tel. +49 41 02 5 64 70
Seite page 137

Heiko Hillig
Dorfstraße 68
D-09518 Mauersberg
Tel. +49 1 71 2 35 54 54
Fax +49 37 35 9 08 62
Seite page 119

Claus Jensen & Henrik Hoelbæk
Tools Design
Rentemestervej 23 A
DK-2400 Kopenhagen NV
Tel. +45 38 19 41 14
Fax +45 38 19 41 13
Seite page 93, 95

Katja Horst
Amalienstraße 15
D-80333 München
Tel. +49 89 28 75 51 31
Fax +49 89 28 75 53 77
mail@KatjaHorst.de
Seite page 147, 151

David Queensbury + Martin Hunt
63 Penfold Street
London NW8 8PQ
Seite page 89

Frank Huster
Eltviller Straße 18
D-65197 Wiesbaden
Tel. +49 6 11 4 73 77
Fax +49 6 11 4 77 47
Seite page 63

Jahn Design Fact
Schlesische Straße 27
D-10997 Berlin
Tel. +49 30 5 30 00 60
Fax +49 30 53 00 06 66
hj@jahn-design.com
www.jahn-design.com
Seite page 33

Hans-Reinhart Janssen
Robert Bosch Hausgeräte GmbH
Robert-Bosch-Straße
D-89537 Giengen
Tel. +49 73 22 92 25 52
Fax +49 73 22 92 25 12
roland.vetter@bshg.com
Seite page 57

Claus Jensen & Henrik Hoelbæk
Tools Design
Rentemestervej 23 A
DK-2400 Kopenhagen NV
Tel. +45 38 19 41 14
Fax +45 38 19 41 13
Seite page 93, 95

Martin Kleinhans
Performa Möbel und Design GmbH
Marbacher Straße 54
D-74385 Pleidelsheim
Tel. +49 71 44 89 88 69
Fax +49 71 44 89 88 76
u.walz@performa.de
www.performa.de
Seite page 39, 135

Matthias Kolb
Obere Rod 3
D-75173 Pforzheim
Mobil +49 1 79 6 77 69 65
ilo-01@gmx.net
Seite page 214

Ralph Krämer
Leberstraße 8
D-66121 Saarbrücken
Tel. +49 6 81 81 79 89
Fax +49 6 81 81 68 53
Seite page 83

Michael Leggewie
Oberer Eschachweg 6
D-89284 Pfaffenhofen
Tel. +49 73 02 31 97
Fax +49 73 02 31 98
design@vision-produktgestaltung.de
Seite page 183

Olavi Lindén
Fiskars Consumer Oy Ab
FIN-10330 Billnäs
Tel. +3 58 19 27 74 23
Fax +3 58 19 23 09 86
o.linden@fiskarsco.fi
Seite page 181

LRG. Licht + Raum AG
Hinterer Schermen 44
CH-3063 Ittingen
Tel. +41 31 9 24 77 88
Fax +41 31 9 24 77 80
LRG@bluewin.ch
Seite page 177

Modus Product Design GmbH
Markgröninger Straße 61/1
D-70435 Stuttgart
Tel. +49 7 11 8 70 21 05
Fax +49 7 11 8 70 21 06
Seite page 163

Clemens Stübner/Sabine Mrasek
Jakobstraße 90
D-52064 Aachen
Tel. +49 2 41 40 02 94 84
Fax +49 2 41 40 02 94 85
Seite page 43

Dr. Silke Nuß, Nuno Rodrigues
Mittnachtstraße 1
D-70191 Stuttgart
Tel. +49 1 60 92 51 55 22
Seite page 31

Ottenwälder und Ottenwälder
Sebaldplatz 6
D-73525 Schwäbisch Gmünd
Tel. +49 71 71 92 72 30
Fax +49 71 71 9 27 23 23
info@ottenwaelder.de
www.ottenwaelder.de
Seite page 111

Markus Paloheimo
Fiskars Consumer Oy Ab
FIN-10330 Billnäs
Tel. +358 19 27 74 27
Fax +358 19 23 09 86
m.paloheimo@fiskarsco.fi
Seite page 181

Ole Palsby
Strandvejen 203
DK-2900 Hellerup
Tel. +45 39 62 00 09
Fax +45 39 62 53 18
Seite page 77

Pearsonlloyd
42–46 New Road
London E1 2AX
Tel. +44 20 73 77 05 60
Fax +44 20 73 77 05 50
Seite page 41

Phoenix Product Design
Kölner Straße 16
D-70376 Stuttgart
Tel. +49 7 11 95 59 76 11
Fax +49 7 11 55 93 92
Seite page 67, 195

Holger Pilz
Schiessmauer 6
D-73116 Wäschenbeuren
pilz-h@hfg-gmuend.de
Seite page 212

Ferdinand A. Porsche
Porsche Design GmbH
Flugplatzstraße 29
A-5700 Zell am See
Tel. +43 65 42 57 22 70
Fax +43 65 42 57 22 72
info@porsche-design.at
Seite page 81

David Queensbury + Martin Hunt
63 Penfold Street
London NW8 8PQ
Seite page 89

Reform Design
Industriestraße 25
D-70565 Stuttgart
Tel. +49 7 11 78 29 53 50
Fax +49 7 11 7 82 95 35 25
info@reform-design.de
www.reform-design.de
Seite page 191

Daniela Breuer, Christian Reichert
Bree Collection GmbH & Co. KG
Gerberstraße 3
D-30916 Isernhagen
Tel. +49 51 36 8 97 60
Fax +49 51 36 8 97 62 29
n.barth@bree.de
www.bree.de
Seite page 123

Werksdesign
Rieber GmbH & Co. KG
Hoffmannstraße 44
D-72770 Reutlingen
Tel. +49 71 21 51 86 40
Fax +49 71 21 51 86 49
info@rieber.de
www.rieber.de
Seite page 187

Frank Rieser, Gerd E. Wilsdorf
Siemens Electrogeräte GmbH
Hochstraße 17
D-81669 München
Tel. +49 89 45 90 23 92
Fax +49 89 45 90 29 58
gerd.wilsdorf@bshg.com
www.siemens-hausgeraete.de
Seite page 51

Dr. Silke Nuß, Nuno Rodrigues
Mittnachtstraße 1
D-70191 Stuttgart
Tel. +49 1 60 92 51 55 22
Seite page 31

Frank Sammet
Sonnenbergstraße 2
D-71723 Großbottwar
Tel. +49 1 71 1 95 33 97
Seite page 179

Rüdiger Schaack
Teichstraße 30
D-08289 Schneeberg
Seite page 29

Roy Schäfer
Weinbergstraße 21
D-75181 Pforzheim
Tel. +49 72 43 33 67 33
Seite page 37

Stefan Schenk
Schüttentobel 15
D-88167 Grünenbach
Mobil +49 1 77 2 30 87 12
schenk-s@hfg-gmuend.de
schuettentobel@hotmail.com
Seite page 212

Steffen Schmidt
Luruper Weg 19
D-20257 Hamburg
Tel. +49 1 60 7 14 14 83
Fax +49 7 11 4 51 01 75 19
Seite page 203

Heike Schnabel, Markus Schneider
Birkenstraße 14
D-73557 Mutlangen
Tel. +49 71 71 77 93 88
Fax +49 71 71 77 93 90
Seite page 165, 167

Alfred Scholze
Alfred Scholze GmbH
Gürtlerstraße 6
D-87600 Kaufbeuren, Neugablonz
Tel. +49 83 41 9 77 00
Fax +49 83 41 6 29 32
Seite page 87

Jürgen Schubert, Thorsten Schwabe,
Peter Zimmer
Mitarbeit: Marcus Kottermair
Blutenburgstraße 41a
D-80636 München
Tel. +49 89 18 95 45 41
Fax +49 89 18 95 45 42
kontakt@thorstenschwabe.de
www.thorstenschwabe.de
Seite page 109

Gregor Baer, Thorsten Schwabe
Indigo Pure Snowboards
Blutenburgstraße 41a
D-80636 München
Tel. +49 89 18 95 45 50
Fax +49 89 18 95 45 49
kontakt@indigosnow.de
www.indigosnow.de
Seite page 107

Mathias Seiler
Kirchstraße 17
D-79539 Lörrach
Tel. +49 76 21 57 77 29
MathiasSeiler@sedus.de
Seite page 131

Josef Sekler
Dalli-Klick
Hasenbergstraße 24
D-73479 Ellwangen-Pfahlheim
Tel. +49 79 65 80 06 32
Fax +49 12 12 5 10 51 79 38
wirsekler@t-online.de
www.wir-einrichtungen.de
Seite page 141

Jeff Shore
Falkoner Allé 44, 4th
DK-2000 Frederiksberg
Tel. +45 40 19 23 35
jpshore@yahoo.com
Seite page 161

Sieger Design
Schloß Harkotten
D-48336 Sassenberg
Tel. +49 54 26 9 49 20
Fax +49 54 26 94 92 39
info@sieger-design.com
www.sieger-design.com
Seite page 65

Boris Simon, Frank Thiele
Factor Product
Comeniusstraße 1
D-81667 München
Tel. +49 89 48 92 78 10
Fax +49 89 48 92 78 11
Seite page 153

SquareOne GmbH
früher:
Topel und Pauser Industrial Design
Talstraße 66
D-40217 Düsseldorf
Tel. +49 2 02 15 92 49 60
Fax +49 2 02 15 92 49 70
info@squareonedesign.de
www.squareonedesign.de
Seite page 113

Staedtler Design Team
Moosäckerstraße 3
D-90427 Nürnberg
Tel. +49 9 11 9 36 50
Seite page 157

Starczewski Design Team
Heimstraße 29
D-89073 Ulm
Tel. +49 7 31 2 80 46
Fax +49 7 31 2 73 80
Seite page 133

Jürgen Steinhauser
3L Modulbausystem
Steinstraße 21
D-90419 Nürnberg
Tel. +49 9 11 77 60 70
Fax +49 9 11 77 60 80
js@3LM.de
www.loftschau.de
Seite page 145

Clemens Stübner/Sabine Mrasek
Jakobstraße 90
D-52064 Aachen
Tel. +49 2 41 40 02 94 84
Fax +49 2 41 40 02 94 85
Seite page 43

T.B.A. Design-Societät
Bismarckstraße 50
D-50672 Köln
Tel. +49 2 21 5 10 20 03
Fax +49 2 21 5 10 30 88
Seite page 47

Thamm Design
Hauptstraße 48
D-74595 Langenburg
Tel. +49 79 05 51 22
Fax +49 79 05 52 01
thamm-design@t-online.de
Seite page 191

Boris Simon, Frank Thiele
Factor Product
Comeniusstraße 1
D-81667 München
Tel. +49 89 48 92 78 10
Fax +49 89 48 92 78 11
Seite page 153

Olaf Tinzmann
Dachziegelwerke Pfleiderer GmbH & Co. KG
Marbacher Straße 4
D-71364 Winnenden
Tel. +49 71 95 69 60
Fax +49 71 95 69 62 25
ranssweiler@pfleiderer-dach.de
www.pfleiderer-dach.de
Seite page 199

Dieter Baumann + Wilfried Trapp
Werksdesign
Baumann + Trapp GbR
Mattenmühle 1
D-77876 Kappelrodeck
Tel. +49 78 41 29 03 00
Fax +49 78 41 29 03 05
ifkka@t-online.de
www.ideen-fuer-kinder.com
Seite page 115, 117

Uno Product Development
Uri Friedländer
20 Belsize Avenue
London, NW3 4AU
Tel. +44 20 77 94 40 80
Fax +44 20 74 31 14 26
Seite page 125

Vistapark GmbH
Villa Media
Viehhofstraße 119
D-42217 Wuppertal
Tel. +49 2 02 2 42 75 00
Fax +49 2 02 2 42 75 80
Seite page 85

Vogt & Weizenegger
Mulackstraße 15
D-10119 Berlin
Tel. +49 30 6 14 21 66
Fax +49 30 6 14 22 87
Seite page 91, 99

Lemonfish
Alexandra Dittrich +
Bettina Weitershagen GbR
Lindenfirststraße 14/1
D-73525 Schwäbisch Gmünd
Tel. +49 71 71 40 51 19
Fax +49 71 71 40 51 24
info@lemonfish.de
www.lemonfish.de
Seite page 121

Frank Rieser, Gerd E. Wilsdorf
Siemens Electrogeräte GmbH
Hochstraße 17
D-81669 München
Tel. +49 89 45 90 23 92
Fax +49 89 45 90 29 58
gerd.wilsdorf@bshg.com
www.siemens-hausgeraete.de
Seite page 51

Johannes Geyer, Gerd E. Wilsdorf
Siemens Electrogeräte GmbH
Hochstraße 17
D-81669 München
Tel. +49 89 45 90 21 59
Fax +49 89 45 90 29 58
gerd.wilsdorf@bshg.com
www.siemens-hausgeraete.de
Seite page 61

Edward Winkelbauer
Industrial Design
Myliusstraße 3
D-71638 Ludwigsburg
Tel. +49 71 41 90 32 22
Fax +49 71 41 92 32 75
Seite page 155

Jürgen Schubert, Thorsten Schwabe,
Peter Zimmer
Mitarbeit: Marcus Kottermair
Blutenburgstraße 41a
D-80636 München
Tel. +49 89 18 95 45 41
Fax +49 89 18 95 45 42
kontakt@thorstenschwabe.de
www.thorstenschwabe.de
Seite page 109

Fotonachweis *Photographs (acknowledgements)*

soweit von den Einreichern erwünscht: *information provided by competitors only:*

Benjamin Gugelberger, Stuttgart
Seite page 142

Reinhold L. Hilgering, München
Seite page 50

Claus Langer, Düsseldorf
Seite page 178

Licht + Schatten, Berlin
Seite page 32

Norbert Mebert, Nürnberg
Seite page 144

Frank M. Orel, Stuttgart
Seite page 24–25, 26–27, 28, 44–45, 46, 70–71, 72, 102–103, 104, 126–127, 128, 148–149, 150, 168–169, 170, 192–193, 194

Susanne Richter
Ö-Konzept, Halle
Seite page 28

Studio Messlinger, München
Seite page 60

Jens Werlein, Schwäbisch Gmünd
Seite page 142

Michael Wolf, Rüsselsheim
Seite page 104

Impressum *Publishing Details*

Ausstellung ***Exhibition***

Veranstalter ***Organizer***

Design Center Stuttgart
des Landesgewerbeamts
Baden-Württemberg
Willi-Bleicher-Str. 19
D-70174 Stuttgart
Tel. +49 711 1 23 26 84

Projektleitung ***Project head***

Sabine Lenk

Organisation ***Administration***

Hildegard Hild
Michael Kern

Ausstellungsgestaltung
Exhibition design

Wolfgang Frank
Landesgewerbeamt
Baden-Württemberg

Fotos zu den Preisträgern, Cover
Photos for the award winners, Cover

Frank Orel
Stuttgart

Ausstellungsdauer
Exhibition dates

10. Oktober 2002 bis
8. Dezember 2002

10 October 2002 to
8 December 2002

Katalog ***Catalogue***

Herausgeber ***Publisher***

Design Center Stuttgart
des Landesgewerbeamts
Baden-Württemberg
Willi-Bleicher-Str. 19
D-70174 Stuttgart
Tel. +49 711 1 23 26 84

Text und Redaktion
Text and editorial supervision

Andrea Scholtz M.A.
Stuttgart

Übersetzung ***Translation***

Philip Mann BA., Ph.D.
Marbach

Grafikdesign ***Graphic design***

Frank Niemeyer
avcommunication AG
Ludwigsburg

Lithografie ***Lithography***

Corinna Rieber
avcommunication AG
Ludwigsburg

Druck ***Printing***

Leibfarth + Schwarz
GmbH + Co. KG
Dettingen/Erms

Produktion ***Production***

Gunther Heeb, Manuela Bloss
avcommunication AG
Ludwigsburg

ISBN
3-929638-67-3
Printed in Germany